Workbook to accompany Essentials of Radiation Biology and Protection

Steve Forshier, MEd, RT(R)

DELMAR
THOMSON LEARNING™

Australia Canada Mexico Singapore Spain United Kingdom United States

Workbook to Accompany
Essentials of Radiation Biology and Protection
by Steve Forshier, MEd, RT(R)

Health Care Publishing Director:
William Brottmiller

Executive Editor:
Cathy L. Esperti

Acquisitions Editor:
Candice Janco

Developmental Editor:
Darcy M. Scelsi

Editorial Assistant:
Maria D'Angelico

Executive Marketing Manager:
Dawn F. Gerrain

Production Editor:
Mary Colleen Liburdi

Printed in United States of America
2 3 4 5 XXX 06 05 04 03 02

For more information contact Delmar, 3 Columbia Circle, PO Box 15015, Albany, NY 12212-5015.

Or you can visit our Internet site at http://www.delmar.com

Library of Congress Cataloging-in-Publication Data on file.

ISBN: 0766813312

Contents

Preface

The *Workbook* to accompany *Essentials of Radiation Biology and Protection* was designed to aid students in the comprehension of critical topics in the field of radiography. Outlines are provided for notes on lecture discussions and review questions offer opportunities to test your knowledge of key terms and core concepts. Each section also contains a practice exam similar to that your instructor may administer. Answers to all questions contained in the workbook are found in the Appendix.

I

Theory and Concepts

History of Radiobiology

OBJECTIVES

Upon completion of this chapter, the reader should be able to:

- Discuss the history of radiobiology
- Identify leaders in the field of radiobiology and their contributions to research
- Define terms related to the measurement of radiation
- Identify regulations related to the field of radiobiology

LECTURE OUTLINE

Radiobiology History

Law of Bergonie and Tribondeau

Ancel and Vitemberger

(continued)

Fractionation Theory

Mutagenesis

Effects of Oxygen

Reproductive Failure

Roentgen

Rad

Rem

____

____

____

Regulation

____

____

____

REVIEW OF KEY TERMS

Match the definition in the right column with the correct term from the left column.

C Direct effect
E Epilation
G Erythema
D Fractionation
F Indirect effect
I ~~J~~ Law of Bergonie and Tribondeau
I Mutagenesis
J ~~I~~ Rad
K Radioactivity
B Radiobiology
A Rem
M Reproductive failure
H Roentgen

a. The unit of dose equivalent or occupational exposure.

b. Division of biology concerned with effects of ionizing radiation on living things.

c. A result of ionization and excitation, an interaction that happens directly on a critical biologic macromolecule.

d. The splitting of radiation into smaller amounts over a period of time.

e. Hair loss.

f. A cell interaction which occurs when the initial ionizing incident takes place on a distant noncritical molecule which then transfers the ionization of energy to another molecule.

g. Reddening of the skin.

h. A unit of radiation exposure descriptive of x- or gamma radiation, the quantity of which would produce a charge.

i. The unit of radiation absorbed dose.

j. States that ionizing radiation is more effective against cells which are highly mitotic, immature, and have long dividing future.

k. The capability of a material to give off rays or particles from its nucleus.

(continued)

l. The causing of genetic mutation by radiation.

m. Cell which is unable to continue repeated divisions after being irradiated.

REVIEW

1. The Law of Bergonie and Tribondeau states:

a. that stem cells are more radiosensitive than mature cells.

b. that older tissues and organs are more radiosensitive than younger tissues and organs.

c. that the lower the metabolic activity of a cell, the more radiosensitive it is.

d. the lower the proliferation and growth rate for tissues, the lower the radiosensitivity.

2. The _______ is the unit of dose equivalent or occupational exposure.

a. roentgen (R)

b. rem

c. rad

3. Which of the following would be considered most radiosensitive?

a. fetus

b. pediatric patient

c. teenage patient

d. adult patient

4. The unit of radiation quantity is the _______.

a. rem

b. rad

c. roentgen

5. Which researcher discovered mutations related to exposure to ionizing radiation?

a. Muller

b. Becquerel

c. Roentgen

d. Curie

EXPLORING THE WEB

1. Search the web for additional information on the history of radiobiology. Create a timeline showing the progress made from the early discoveries to today.

2. Search the web for current research being done in the area of radiobiology. What new discoveries or theories are on the horizon? What impact may these new discoveries have on current practice?

3. Search the web for regulations in the field of radiobiology. Find the web sites of agencies responsible for enforcing and revising regulations. Are there any new regulations pending? What are the penalties for failing to meet regulations? What additional information can you find pertaining to regulation in the field of radiobiology?

Cellular Anatomy and Physiology

OBJECTIVES

Upon completion of this chapter, the reader should be able to:

- Indicate the parts of the cell
- Identify organic compounds and their functions
- Identify inorganic compounds and their functions
- Explain mitosis
- Explain meiosis

LECTURE OUTLINE

Cell Biology

____

____

____

Chemical Configuration of Cells

____

____

____

Cell Structure

____

____

____

(continued)

Cell Growth and Division

____

____

____

Mitosis

____

____

____

Meiosis

____

____

____

REVIEW OF KEY TERMS

Match the definition in the right column with the correct term from the left column.

___ Amino acid

___ Anabolism

___ Anaphase

___ Autosome

___ Catabolism

___ Centromere

___ Chromatid

___ Chromatin

___ Chromosome

___ Deoxyribonucleic acid (DNA)

___ Diploid

___ Duplication

a. RNA that carries amino acids to ribosomes for assisting in protein synthesis.

b. Having half the diploid number of chromosomes found in somatic cells.

c. An organic compound which is the building block of proteins, and the end product of protein digestion.

d. Any chromosome that is other than the sex chromosome.

e. The gap or growth period between telophase and the start of DNA synthesis when the DNA is not replicating.

f. The gap or growth period following the replication of DNA and prior to mitosis.

g. The stage of mitosis where chromosomes are arranged.

___ Enzyme

___ G_1

___ G_2

___ Gamete

___ Gene

___ Haploid

___ Interphase

___ Macromolecule

___ Meiosis

___ Messenger RNA

___ Metaphase

___ Mitosis

___ Nuclear membrane

___ Nucleolus

___ Organism

___ Polymer

___ Prophase

___ Protoplasm

___ Ribonucleic acid (RNA)

___ Ribosomal RNA

___ S-phase

___ Telophase

___ Transfer RNA

h. A molecule created by combining two or more of the same molecules.

i. Period of synthesis or replication.

j. The construction phase of metabolism, when a cell takes the substances from blood that are necessary for repair and growth, and converts them into cytoplasm.

k. The constricted area of the chromosome that separates the chromosome into two arms.

l. Possessing two sets of chromosomes.

m. The basic unit of heredity which has a specific location on chromosome.

n. A two-layered membrane which surrounds the cell nucleus.

o. Spherical body in the cell nucleus that holds nuclear RNA.

p. Any living entitiy.

q. The final stage of mitosis or meiosis, during which there is reconstruction of the nuclear membrane, and cell cytoplasm divides, giving birth to two daughter cells.

r. Nucleic acid which controls protein synthesis.

s. Type of cell division involving somatic cells in which a parent cell divides to create two daughter cells that contain the same chromosome number and DNA content as the parent.

t. The period between cell division, known as the resting stage, when DNA is being synthesized.

u. The linear thread of a cell nucleus.

v. A stage in mitosis and meiosis between metaphase and telophase in which chromatids migrate toward opposite poles of the cell.

w. The destructive phase of metabolism, in which complex substances are changed into simpler substances.

x. The halves into which the chromosome is longitudinally divided, which are held together by the centromere and move to opposite poles of a dividing cell during anaphase.

y. A material in the nucleus which contains genetic information.

z. A polymer composed of deoxyribonucleotides, arranged in a double helix.

(continued)

aa. A chromosome mutation in which either one or both segments of a chromosome join to another chromosome.

bb. A complex organic protein which accelerates chemical reactions.

cc. The mature male or female reproductive cell.

dd. A large molecule.

ee. Cell division of germ cells, which consists of two cell divisions but only one replication of DNA.

ff. RNA that binds amino acids to ribosomes during protein synthesis.

gg. The first stage of mitosis or meiosis when chromosomes become visible.

hh. A colloidal structure of organic and inorganic materials and water which form the living cell.

ii. RNA that exists in ribosomes and assists in protein synthesis.

REVIEW

1. A combination of two or more tissues which are combined to perform a specific function is a definition for:

a. cells

b. tissues

c. organs

d. systems

2. _______________ assist in growth, construct new tissues, and repair injured or worn-out cells.

a. Proteins

b. Enzymes

c. Lipids

d. Carbohydrates

3. The location of genetic information is in the:

a. cell membrane.

b. endoplasmic reticulum

c. mitochondria

d. nucleus

4. Which of the following is the RNA nucleotide base that pairs with adenine in DNA synthesis?

a. thymine

b. uracil

c. guanine

d. cytosine

5. The normal diploid or 2n number in humans is:

a. 11

b. 23

c. 46

d. 46 pair

6. In which phase of mitosis do the chromosomes line up at the equator of the cell?

a. prophase

b. metaphase

c. anaphase

d. telophase

7. During meiosis, or reduction division,
   a. the cell divides twice in succession, but chromosomes are duplicated only one time
   b. the cell divides twice in succession, and chromosomes are duplicated twice
   c. the cell divides only once, and chromosomes are duplicated only one time
   d. the cell divides only once, but chromosomes are duplicated twice

EXPLORING THE WEB

1. Search the web using the key terms mitosis and meiosis. Can you find any demonstrations of these concepts?
2. Search the web for organic compounds. Describe the roles and functions of these compounds in the human body.
3. Search the web for inorganic compounds. Describe the roles and functions of these substances in the human body.

LABELING

Label the parts of the cell.

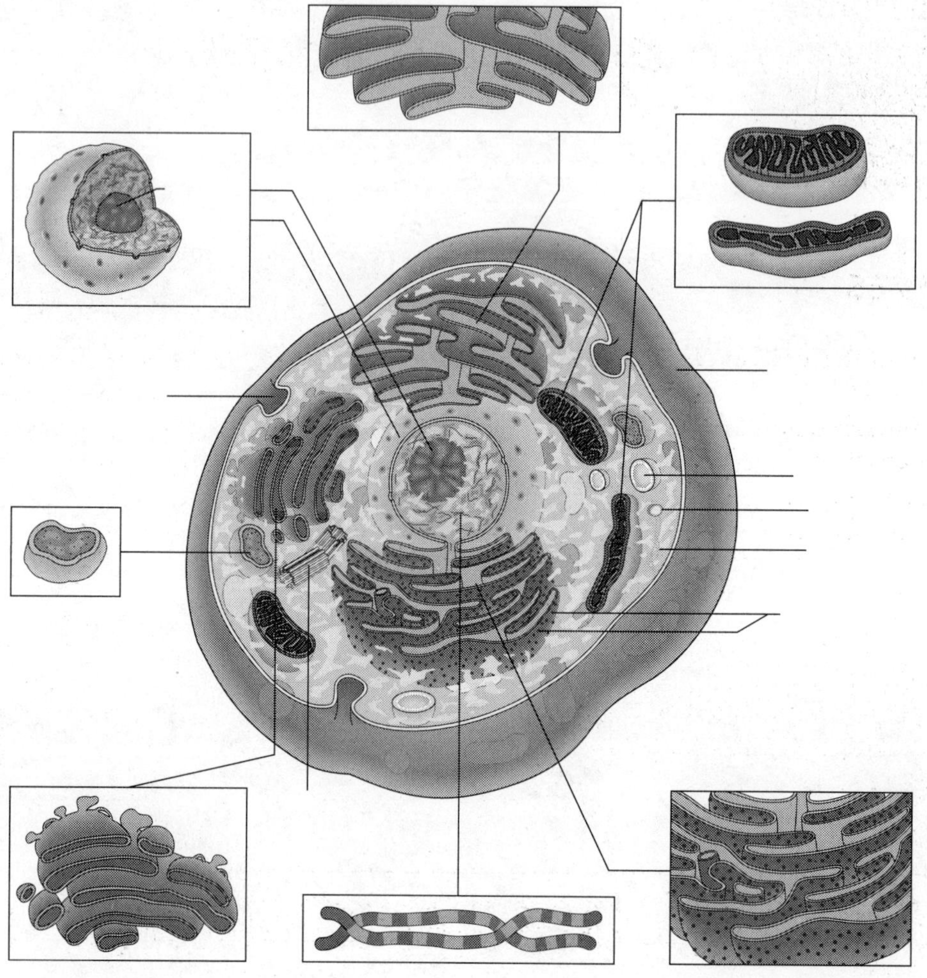

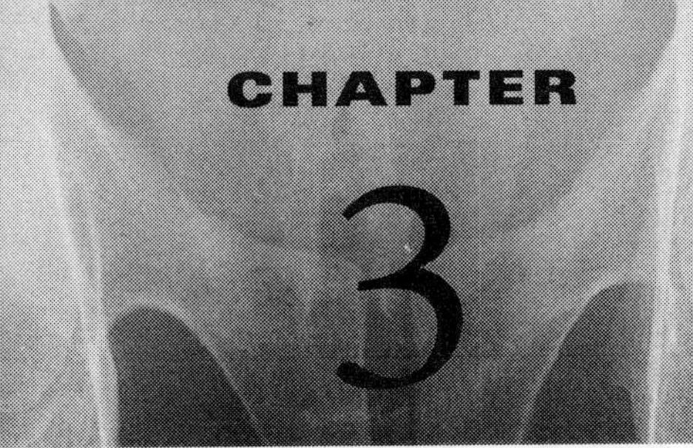

Cellular Effects of Radiation

OBJECTIVES

Upon completion of this chapter, the reader should be able to:

- Examine the physical and biologic factors affecting cell radiosensitivity
- Inspect the direct and indirect effects of radiation
- Evaluate the radiolysis of water
- Explain the irradiation of macromolecules
- Analyze the types of dose-response relationships
- Discuss target theory
- Explain cell survival curves

LECTURE OUTLINE

Radiosensitivity of Cells

___

___

___

Physical and Biologic Factors

___

___

___

Direct and Indirect Effects of Radiation

___

___

___

(continued)

Interactions with Radiation

Radiolysis of Water

Irradiation of Macromolecules

Single-hit Chromosome Aberrations

Multi-hit Chromosome Aberrations

Reciprocal Translocations

Dose-Response Relationships

Linear Dose-Response Relationships

Linear Quadratic Dose-Response Curves

Target Theory

Cell Survival Curves

REVIEW OF KEY TERMS

Match the definition in the right column with the correct term from the left column.

___ Deletion

___ Dicentric

___ Division delay

___ Dose-response relationship (curve)

___ Effectiveness

___ Free radical

___ HeLa

___ Interphase death

___ Linear dose-response curve

___ Linear energy transfer (LET)

___ Linear quadratic dose-response curve

___ Mutation

___ Nonthreshold

___ Oxygen effect

___ Point mutation

___ Polyribosomes

___ Radiolysis

___ Radiosensitivity

___ Relative biolgic effectiveness (RBE)

___ Ring

___ Sigmoid

___ System

___ Threshold

___ Translocation

a. The rate at which a desired effect is produced.

b. A chromosomal effect that causes a loss of genetic material.

c. An S-shaped type of dose-response relationship.

d. Altering of a chromosome either by a portion of it transferring to another chromosome or to another section of the same chromosome.

e. A chromosome which has two centers or two centromeres.

f. A measure of the rate at which energy is deposited from ionizing radiation to soft tissue.

g. A structural change or transformation of a chromosome which can be transmitted to offspring.

h. Breakdown of water using radiation.

i. A group of cells that perform a particular function.

j. The point where a stimulus starts to produce an effect.

k. A mutation which occurs as a result from a change of a single DNA base pair, created by one nucleotide being exchanged for another.

l. A graphical representation of observed effects compared with radiation dose.

m. Cell death occurring before mitosis.

n. The amount of reaction or response of a cell to radiation.

o. Mutation of chromosome causing it to become ring-shaped.

p. A comparison of how effective types of radiation are compared with x- and gamma-rays.

q. Name given to cell when response to radiation when oxygen is present.

r. Any dose received, regardless of size, which will produce a response.

s. An atom which has an unpaired electron making it highly reactive.

t. Graphical representation of the relationship between radiation dose and observed response, in which any dose may have a potential effect, and there is a direct relationship between radiation dose and observed effect.

u. Letters which represent the first two letters of the patient's first and last names in an experiment conducted by Puck using uterine cervix cells to produce cell cultures.

v. Graphical representation of relationship between radiation dose and observed response, in which the curve is linear or proportional at low doses, and becomes curvilinear at higher doses.

REVIEW

1. Which of the following cell groups are considered highly radiosensitive?

a. lymphocytes, spermatogonia, erythroblasts, and intestinal crypt cells

b. endothelial cells, osteoblasts, spermatids, and fibroblasts

c. muscle and nerve cells, and chondrocytes

2. Relative biologic effectiveness (RBE):

a. describes a measure of the rate at which energy is deposited as a charged particle travels through matter

b. is a comparison of a dose of test radiation to a dose of 250-keV x-ray that produces the same biologic response

c. is defined as the dose of radiation that produces a given biologic response under anoxic conditions divided by the dose of radiation that produces the same biologic response under aerobic conditions

3. Which of the following body molecules are most commonly acted upon directly by ionizing radiation to produce indirect effects?

a. lipids

b. proteins

c. carbohydrates

d. water

4. The types of DNA molecule damage include:

a. main-chain scission, one side rail broken

b. main-chain scission, both side rails broken

c. main-chain scission, resulting in cross-linking

d. rung breakage, causing bases to separate

e. a changing or loss of a base

5. At low doses of radiation, most cellular radiation damage is the result of:

a. main-chain scission

b. cross-linking

c. point lesions

6. Nonthreshold:

a. means that an observed response is directly proportional to the dose

b. means that an observed response is not directly proportional to the dose.

c. assumes that there is a radiation level reached under which there would be no effects observed

d. assumes that any radiation dose produces an effect

7. According to target theory:
   a. only direct effects can cause hits
   b. only indirect effects can cause hits
   c. RNA is considered the critical target
   d. DNA is considered the critical target

8. Which portion of the cell survival curve proves that some damage must accumulate before cell death can occur?
   a. shoulder
   b. straight-line
   c. toe

EXPLORING THE WEB

1. Search the web for information related to dose-response curves and radiobiology. What additional information can you find?

2. Search for effects of radiation at the cellular level. Describe the impact radiation has in cells. Discuss the effects of cellular radiation as a treatment for cancer.

3. Search for radiation therapy. Discuss the concepts of radiation therapy and relate them to the discussion in your text.

4. Search for more information on the target theory and the oxygen effect.

SECTION I SELF-ASSESSMENT QUESTIONS

1. The process of cell division of somatic cells is known as:
   a. mitosis
   b. synthesis
   c. meiosis
   d. reduction division

2. Of the following stages of mitosis, which is considered the most radiosensitive?
   a. prophase
   b. anaphase
   c. metaphase
   d. telophase

3. Which of the following describes the shape of a DNA molecule?
   a. oval
   b. spherical
   c. rectangular
   d. double helix

4. How many matched pairs of chromosomes does a normal human cell contain?
   a. 11
   b. 23
   c. 46
   d. 47

5. Which stage of cell division is also known as the "resting stage"?
   a. prophase
   b. anaphase
   c. metaphase
   d. interphase

6. If a DNA base sequence is altered, which of the following would occur?
   a. a gene mutation
   b. a gene duplication
   c. a gene replication
   d. formation of chromatin

7. Of the following, which are considered the "building blocks" of protein synthesis?
   a. chromosomes
   b. genes
   c. amino acids
   d. lipids

8. In which area of the cell is the majority of RNA located?
   a. nucleolus
   b. mitochondria
   c. lysosomes
   d. cytoplasm

9. The small areas of the DNA molecule that determine cell characteristics are named:
   a. adenines
   b. pyrimidines
   c. genes
   d. nucleolus

10. How does oxygen retention affect cell radiosensitivity?
   a. increases radiosensitivity
   b. decreases radiosensitivity
   c. destroys radiosensitivity
   d. does not affect radiosensitivity

11. What is the name of the molecule that has one or more unpaired electrons and is usually chemically reactive?
   a. centromere
   b. polypeptide
   c. free radical
   d. ion

12. Which of the following contains the human hereditary blueprint?
   a. the RNA
   b. the nucleolus
   c. the gene
   d. the ribosome

13. In mitosis, chromosomes split in half. What are these two halves called?
   a. nuclear membranes
   b. chromatids
   c. centromeres
   d. spindle fibers

14. For protein synthesis to occur, messenger RNA (m-RNA) carries information to which of the following?
   a. DNA
   b. t-RNA
   c. chromosome
   d. ribosome

15. Transfer RNA (t-RNA) carries which of the following in order to synthesize protein?
   a. monosaccharides
   b. polysaccharides
   c. genes
   d. amino acids

16. The majority of a cell's genetic information is found where?
   a. nucleus
   b. cytoplasm
   c. t-RNA
   d. nucleolus

17. The process of cell division of reproductive/germ cells is named:
   a. mitosis
   b. meiosis
   c. synthesis
   d. transcription

18. The following is in reference to the single set of chromosomes in a germ cell:
   a. tetrad
   b. haploid number
   c. diploid number
   d. tetroid number

19. Which of the following measures the rate of energy lost along the track of an ionizing particle?
   a. relative biologic effectiveness (RBE)
   b. linear energy transfer (LET)
   c. oxygen enhancement ratio (OER)
   d. cell survival curve

20. What does the term interphase death mean?
   a. cells die prior to entering interphase
   b. cells die prior to leaving interphase
   c. cells die in between mitotic phases
   d. the organism dies

21. Which of the following terms is used in describing cell damage from radiation that is not sufficient to kill the cell?
   a. sublethal damage
   b. duplication effect
   c. cell division
   d. replication

22. Why is free radical formation considered such a threat to humans?
   a. free radicals produce scatter radiation
   b. free radicals can penetrate any type of shielding
   c. free radicals have been proven to have carcinogenic effects
   d. free radicals have been observed to produce toxic effects

23. According to the target theory, which of the following is thought to be the principal target of cell damage?

a. RNA

b. DNA

c. cytoplasm

d. ribosomes

24. What does the term "indirect effect" of ionizing radiation refer to?

a. genetic effects are produced

b. an ionization occurs directly on the target molecule

c. an ionization occurs in one location which can produce effects at a distant location

d. organism death occurs

25. Which of the following would be considered most radiosensitive?

a. a fetus

b. a pediatric patient

c. a teenaged patient

d. an adult patient

26. Which of the following is the term used to describe the separation of water into hydrogen and oxygen following irradiation?

a. duplication

b. synthesis

c. radioactivity

d. radiolysis

27. Which of the following is used for expressing occupational exposure?

a. rem

b. rad

c. roentgen

d. LET

28. Of the stages of mitosis, which is the most radiosensitive?

a. prophase

b. anaphase

c. metaphase

d. telophase

29. As linear energy transfer (LET) increases, how is relative biologic effectiveness (RBE) affected?

a. with an increase in LET, RBE also increases

b. with an increase in LET, RBE decreases

c. with an increase in LET, RBE is neutralized

d. RBE is not affected by LET

30. To correctly make a DNA base pair, guanine must be bonded to:

a. thymine

b. cytosine

c. guanine

d. adenine

31. The type of irradiation damage most likely to cause abnormalities in base sequences, and thus cell mutation, would be

a. single-strand breaks

b. double-strand breaks

c. cross-linking

d. base damage

SECTION

II

Biological Effects of Radiation Exposure

CHAPTER 4

Effects of Initial Exposure to Radiation

OBJECTIVES

Upon completion of this chapter, the reader should be able to:

- Discuss the hematologic, gastrointestinal, and central nervous system syndromes
- Describe local tissue damage to the skin, eyes, and gonads
- Explain hematologic and cytogenic effects

LECTURE OUTLINE

Acute Radiation Syndromes

Response Stages

Bone Marrow Syndrome

(continued)

Gastrointestinal Syndrome

Central Nervous System Syndrome

Local Tissue Damage

Skin

Eyes

Gonads

Hematologic Effects

____

____

____

Hemopoietic System

____

____

____

Cytogenic Effects

____

____

____

REVIEW OF KEY TERMS

Match the definition in the right column with the correct term from the left column.

___ Aberration

___ Acentric

___ Alopecia

___ Anemia

___ Anomaly

___ Atrophy

___ Crypts of Lieberkuhn

___ Cytopenia

___ Dermis

___ Desquamation

___ Dicentric

___ Edema

a. Imperfection.

b. Swelling.

c. Open wound.

d. Inflammation of blood vessels.

e. The relationship of signs and symptoms to a specific disease or trauma.

f. On the outside.

g. Peeling skin.

h. The second stage in the response to radiation exposure in which changes are taking place within the body system that may either result in death or recovery.

i. Radiosensitive cells precursor to the population of villi cells.

___ Epidermis

___ Erythroblasts

___ Follicles

___ Granulocytes

___ Hemopoietic

___ Inflammation

___ Karyotype

___ Latent

___ LD50/30

___ LD50/60

___ Lesion

___ Manifest

___ Maturation depletion

___ Megakaryocytes

___ Meningitis

___ Myelocytes

___ Necrosis

___ Oogonia

___ Prodromal

___ Radiation cataractogenesis

___ Radioresistant

___ Sebaceous

___ SED50

___ Skin erythema dose (SED)

___ Somatic

___ Spermatogonia

___ Stem cell

___ Stroma

___ Subcutaneous

___ Syndrome

___ Threshold dose

___ Vasculitis

j. Immature cells.

k. Radiation dose necessary to affect 50% of a population.

l. The third stage in the response to radiation exposure in which body systems show signs and symptoms of exposure.

m. Reduction in red blood cell counts.

n. Shrinking.

o. Departure from normal.

p. Depression of all blood cell counts.

q. Precursors for red blood cells.

r. The dose at which symptoms will occur.

s. Supportive tissue of an organism.

t. Red blood cells with a life cycle of one day.

u. A measure of the amount of radiation a person received.

v. Pertaining to the development of blood cells.

w. Having no reaction or ill effects of exposure to radiation.

x. Chromosome map.

y. The first stage in the response to radiation exposure.

z. Formation of cataracts caused by exposure to radiation.

aa. Hair loss.

bb. Middle layer of the skin.

cc. A chromosome which has two centers or two centromeres.

dd. Outer layer of the skin.

ee. Sacs or cavities.

ff. Inflammation of the membranes of the spinal cord and brain.

gg. Redness and swelling at site of exposure.

hh. Precursors for white blood cells.

ii. Lethal dose to kill 50% of a population in 30 days.

jj. Tissue death.

kk. Lethal dose to kill 50% of a population in 60 days.

ll. Reproductive cells of the female.

mm. Precursors for platelets.

nn. Reduction in the number of mature sperm.

oo. Nonreproductive cells.

pp. Oily secretion.

qq. Reproductive cells of the male.

rr. Inner layer of skin.

REVIEW

1. What three conditions must be met in order for the total body radiation syndrome to be applicable?

2. What is the primary result of an acute radiation exposure?

3. What is the approximate value of LD50/30 in humans?

4. Name the three acute radiation syndromes and their approximate dose ranges.

5. List in correct order the response stages of disease as they occur.

6. What is the cause of death in the bone marrow syndrome?

7. What is the cause of death in the GI syndrome?

8. What is thought to be the cause of death in the CNS syndrome?

9. Explain the SED50.

10. What dose-response relationship do radiation-induced cataracts follow?

11. What is the main effect of irradiation to the male gonads?

12. What doses are necessary to cause sterility in the male and female?

13. Explain the major difference in radiosensitivity between the testes and ovaries.

14. List the bone marrow cells in correct order from least sensitive to most sensitive.

15. What are the consequences of cell depletion as discussed above?

16. Explain the difference between chromosome aberrations and chromatid aberrations.

17. List the types of single- and double-break chromosome aberrations which are possible.

18. Explain the difference between somatic and genetic mutations.

19. Explain the data which are available from radiation cytogenetic studies.

EXPLORING THE WEB

1. Search the web for additional information on acute radiation syndromes. What resources are available? Create flash cards outlining the signs and symptoms of each syndrome for review and study.

2. Search the web for examples of local tissue damage related to radiation exposure. What are you able to find on radiation exposure of skin, eyes, and gonads? How are females affected differently than males?

3. Search the web for examples of chromosomal and cellular effects of radiation exposure. Can you find any case studies that illustrate these effects? What additional information can you find on the effects of radiation exposure on the hematologic system?

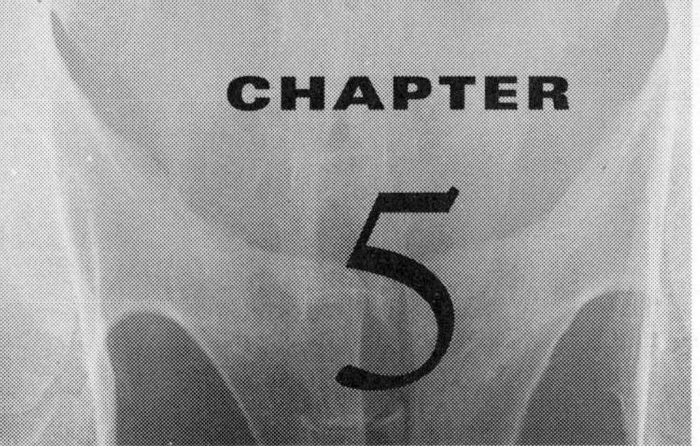

Effects of Long-term Exposure to Radiation

OBJECTIVES

Upon completion of this chapter, the reader should be able to:

- Discuss epidemiology
- Explain risk estimation models
- Examine radiation-induced malignancies
- Identify life span shortening
- Discuss genetic damage
- Explain irradiation of the fetus
- Analyze stochastic and nonstochastic effects

LECTURE OUTLINE

Epidemiology

____

____

____

Dose-Response Curves

____

____

____

Relative Versus Absolute Risk

____

____

____

(continued)

Radiation induced Malignancies

Leukemia

Skin Carcinoma

Thyroid Cancer

Breast Cancer

Osteosarcoma

Lung Cancer

Life Span Shortening

Genetic Damage

Irradiation of the Fetus

Pre-implantation Stage

Fetal Growth Stage

(continued)

Stochastic and Nonstochastic Effects

____

____

____

REVIEW OF KEY TERMS

Match the term in the left column with the correct definition from the right column.

___ Absolute risk model

___ Acute

___ Ankylosing spondylitis

___ BEIR

___ Benign

___ Carcinogenic

___ Carcinoma

___ Chronic

___ Doubling dose

___ Excess risk

___ Fibrosis

___ Follicular

___ Genetically significant dose (GSD)

___ Hypoxic

___ Leukemia

___ Loci

___ Lymphoid

___ Malignant

___ Myeloid

___ Neoplasm

___ Neuroblasts

___ Nonstochastic

___ Osteosarcoma

___ Papillary

a. Tumor.

b. Marrow.

c. Deterministic.

d. An endocrine disorder caused by the failure of the ovaries to respond to pituitary hormone stimulation.

e. Occurring randomly in nature.

f. The ratio of cancer incidence in an exposed population to that of an unexposed population.

g. An average calculated from the gonadal dose received by the entire population and used to determine the genetic influence of low dose to whole population.

h. Low levels of oxygen.

i. Cancer of blood forming cells in bone marrow.

j. Essential life sustaining cells.

k. Offspring.

l. Radiation producing cancer.

m. Radiogenic substance.

n. Radiogenic gas.

o. Committee on the Biological Effects of Ionizing Radiation.

p. Estimates a continual increase in risk, independent of the age-specific cancer risk at time of exposure.

q. Cancer causing.

r. A growth or tumor.

___ Parenchymal

___ Progeny

___ Radiocarcinogenesis

___ Radium

___ Radon

___ Relative risk model

___ Stochastic

___ Turner's syndrome

___ Zygote

s. The dose of radiation required per generation to double the spontaneous mutation rate.

t. The number of excess cases of cancer observed compared with expected spontaneous occurrence.

u. Nonprogressive.

v. Cavity.

w. Spot, place of origin.

x. Bone cancer.

y. Immobility of the vertebrae.

z. Fertilized egg.

aa. Abnormal formation of fibrous tissue.

bb. Rapid, severe.

cc. Lymph tissue.

dd. Progressive; threatening.

ee. Embryonic nerve cells.

ff. Nipple-like protrusion.

gg. Slow, progressive.

REVIEW

1. List the four population groups from whom we obtain data on the incidence of radiation-induced cancer.

____

____

____

2. Explain the difference between the relative (multiplicative) and absolute (additive) risk models.

____

____

____

3. Explain the difference between relative, absolute, and excess risk.

____

____

____

4. Using the previously mentioned data, calculate the relative risk for radiation-induced leukemia for the Hiroshima and Nagasaki atomic bomb survivors.

____

____

____

(continued)

5. Why were the incidences of radiation-induced cancer at low doses different between Hiroshima and Nagasaki?

6. List the shape of the radiation dose-response relationship, the latent period for, and the at-risk period for radiation-induced leukemia.

7. List the dose-response relationship and latent period for radiation-induced skin cancer.

8. Approximately what percent of deaths are attributed to radiation-induced thyroid cancer? Why do females have a greater risk? What dose response relationship does radiation-induced thyroid cancer follow?

9. What dose-response relationship does radiation-induced breast cancer follow? What is the latent period? What is the absolute risk?

10. According to data obtained from the watch-dial painters, what is the overall relative risk of osteosarcoma? What is the absolute risk? What dose-response relationship does osteosarcoma follow?

11. From data obtained from studies on miners, what dose-response relationship does radiation-induced lung cancer follow? What is the absolute risk?

12. What evidence is available to indicate life span shortening in humans?

13. What data were obtained from Muller's fruit fly experiments and the Russell mice experiment? How do the data differ?

14. What factors are calculated in the genetically significant dose (GSD)?

15. Explain the concept of doubling dose.

16. How does the BEIR Committee view radiation-induced chromosome aberrations and recessive mutations?

17. What four factors influence the embryo's response to irradiation?

18. Explain the all-or-nothing response.

19. What do data obtained from animal experiments tell us about the pre-implantation stage?

20. Why is the development of the CNS in humans important when discussing irradiation of the fetus?

21. What are the three stages of the human gestational period?

22. Explain the data that were obtained from the atomic bomb survivors.

23. What is the predominant scientific opinion regarding radiation-induced congenital abnormalities?

24. What are the current mortality estimates for radiation-induced leukemia?

25. What is the current effective dose-equivalent limit to the fetus during the gestational period?

26. What precautions should be taken to protect the fetus from possible irradiation?

(continued)

27. Contrast stochastic and nonstochastic effects.

____

____

____

EXPLORING THE WEB

1. Search the web for sites discussing the victims of Hiroshima, Nagaskai, and Chernobyl. What types of injury and mutation were discovered in these populations? What were the rates of morbidity and mortality in these areas?

2. Search the web for additional information on risk models related to radiation exposure. Are there additional models that were not discussed in the text? How do they relate to what you have already learned? Can you find any case studies related to assessment of risk related to radiation exposure?

3. Search the web for additional information on radiation-induced malignancies. What are the morbidity and mortality rates associated with each type of cancer found? Are incidences higher in males than in females?

SECTION II SELF-ASSESSMENT QUESTIONS

1. For humans, the LD50/30 is approximately
   a. 50 rad
   b. 100 rad
   c. 200 rad
   d. 300 rad

2. Pertaining to irradiation of mammalian gonads,
   a. effects are independent of LET
   b. for sterility, the dose-response relationship is linear, nonthreshold
   c. in the male the spermatocyte is more radiosensitive than the spermatogonia
   d. depression of germ cells has been measured at doses as low as 10 rad

3. The LD50/30 is representative of the dose
   a. necessary to kill 10% of the cells in 50 days
   b. necessary to kill 50% of the cells in 50 days
   c. necessary to kill 50% of the cells in 30 days
   d. necessary to kill 30% of the cells in 50 days

4. Which of the following would be considered an early response to irradiation?
   a. genetic damage
   b. leukemia
   c. life span shortening
   d. cytogenetic damage

5. The type of dose-response relationship demonstrated by human radiation lethality is
   a. linear, threshold
   b. nonlinear, threshold
   c. nonlinear, nonthreshold
   d. linear, nonthreshold

6. What is the approximate dose necessary to the ovaries to produce permanent sterility?
   a. 50 rad
   b. 100 rad
   c. 150 rad
   d. 200 rad

7. Which acute radiation syndrome has a mean survival time that is independent of dose?

a. hematologic

b. gastrointestinal

c. central nervous system

8. What is the approximate human SED50?

a. 100 rad

b. 300 rad

c. 600 rad

d. 1000 rad

9. Which of the following is *not* an early response to irradiation?

a. breast cancer

b. skin erythema occurring 2 weeks postexposure

c. intestinal distress occurring 1 week postexposure

d. chromosome aberrations

10. Which of the following is *not* an acute local tissue effect of radiation exposure?

a. cataracts

b. epilation

c. temporary sterility

d. skin erythema

11. Death from to a single dose of whole body irradiation primarily involves damage to the

a. skin

b. bone marrow

c. skeletal system

d. respiratory system

12. A whole body radiation dose given in a period of seconds to minutes produces a clinical pattern called

a. mortality rate

b. relative biological effectiveness

c. acute radiation syndrome

d. clinical body dose

13. What does "GSD" represent?

a. general safe dose

b. gonad safe dose

c. germ safe dose

d. genetically significant dose

14. What is the minimum radiation level below which no genetic or somatic damage would occur?

a. 1 R

b. 5 R

c. 10 R

d. no minimum level exists

15. When does the greatest radiation hazard to a fetus exist?

a. first trimester

b. second trimester

c. third trimester

16. What is the principal response of the blood caused by radiation exposure?

a. chromosome rearrangement

b. chromosome fragmentation

c. decrease in cell number

d. cell proliferation stimulation

17. A reddening of the skin caused by radiation damage is referred to as

a. epistaxis

b. epilation

c. erythema

d. cataractogenesis

18. Radiation cataractogenesis

a. follows a linear, nonthreshold dose-response relationship

b. follows a nonlinear, threshold dose-response relationship

c. exhibits a threshold of approximately 5 rad

d. has a latent period of 6 months

19. Epidemiology is defined as the study of

a. populations

b. radiation

c. physics

d. statistics

20. The relative risk for development of radiation-induced leukemia is approximately

a. 0.1

b. 0.5

c. 1.0

d. 1.5

21. The absolute risk factor for radiation-induced breast cancer is approximately ___ cases/10^6/rad/year.

a. 0.06

b. 1.0

c. 6

d. 60

22. Radiation-induced lung cancer exhibits which of the following dose-response relationships?

a. nonlinear, nonthreshold

b. linear, nonthreshold

c. linear, threshold

d. nonlinear, threshold

23. What is the approximate latent period for radiation-induced leukemia?

a. 6 months

b. 1 year

c. 4 to 7 years

d. 20 years

24. Which of the following populations has *not* shown an increased incidence of radiation-induced leukemia?

a. American radiographers

b. American radiologists

c. atomic bomb survivors

d. radiotherapy patients

25. Which of the following types of radiation-induced cancers demonstrates a threshold dose-response relationship?

a. skin

b. bone

c. lung

d. breast

26. Which of the following is most likely after irradiation in utero if the radiation is received during the organogenesis stage?

a. prenatal death

b. neonatal death

c. latent malignancies

d. congenital abnormalities

27. What is the term which describes radiation damage that increases the probability of causing a late effect but will not increase the severity of the effect?

a. chronic

b. congenital

c. stochastic

d. nonstochastic

28. Which of the following is the term used to describe the dose of radiation which will increase the number of mutations by a factor of two?

a. doubling dose

b. stochastic dose

c. nonstochastic dose

d. congenital dose

29. What is the approximate dose required to cause permanent male sterility?

a. 5 rad

b. 100 rad

c. 250 rad

d. 500 rad

30. Which of the following radiation syndromes would be caused by an acute whole body exposure of 5,000 rad?

a. hematologic syndrome

b. gastrointestinal syndrome

c. central nervous system syndrome

31. Which of the following describe ulceration, necrosis, and loss of skin cells from irradiation?

a. erythema

b. epilation

c. desquamation

32. Which of the following is considered the most radiosensitive?

a. fetus

b. pediatric

c. adult

d. geriatric

33. The annual dose-equivalent limit for the fetus of an occupational worker is

a. 0.001 rem

b. 0.01 rem

c. 0.5 rem

d. 1.0 rem

SECTION

III

Radiation Protection

Protection of Personnel

OBJECTIVES

Upon completion of this chapter, the reader should be able to:

- Discuss the rationale for radiation protection
- Explain personnel dosimeters, dosimetry reports, and duties of the radiation safety officer
- Define and calculate the dose-limiting recommendations for diagnostic radiology personnel
- Explain structural shielding construction and list the items that influence this construction
- Describe how to decrease the radiographer's exposure during a mobile radiographic examination
- Identify ways to lesson the radiographer's exposure during a flouroscopic examination
- Discuss how using distance can decrease radiation exposure
- Illustrate the inverse square law
- Identify garments that can be worn to reduce radiation exposure and explain when such garments should be used
- List the people and methods that can help with patient immobilization during an x-ray exposure

LECTURE OUTLINE

Rationale for Radiation Protection

___

___

___

Monitoring of Personnel

___

___

___

(continued)

Film Badges

Thermoluminescent Dosimeters

Pocket Dosimeters

Dosimetry Report

Dose-limiting Recommendations

Principles of Personnel Radiation Exposure

Time

Distance

Shielding

Structural Shielding Construction

Use of Protective Garments

Mobile Exam Considerations

(continued)

Flouroscopic Exam Considerations

____

____

____

Inverse Square Law

____

____

____

Patient Immobilization Considerations

____

____

____

REVIEW OF KEY TERMS

Match the definition in the right column with the correct term from the left column.

___ Agreement states

___ ALARA

___ Bone marrow dose

___ Controlled area

___ Cumulative timing device

___ Deadman type

___ Dose

___ Dosimetry

___ Early effect of radiation

___ Effective dose-equivalent limit

___ Entrance skin exposure (ESE)

___ Film badge

a. The amount of activity of the x-ray machinery.

b. The primary x-ray beam.

c. A fixed barrier that is located parallel to the line of travel of the primary x-ray beam.

d. Radiation that is dissipated away from the point of origin.

e. Methods used to limit exposure to radiation.

f. A fixed barrier that is located perpendicular to the line of travel of the primary x-ray beam.

g. The amount of time a hospital area is occupied by people.

h. The amount of material that is needed to meet the same requirements as a lead barrier.

i. A tube containing a positive and negative electrode that are charged before use. One

___ Gonadal dose

___ Intermittent fluoroscopy

___ Inverse square law

___ Ionization chamber

___ Joint Commission on the Accreditation of Healthcare Organizations (JCAHO)

___ Late effect of radiation

___ Lead equivalent

___ Personnel dosimeters

___ Packet dosimeter

___ Primary protective barrier

___ Protective tube housing

___ Radiation protection

___ Scattered radiation

___ Secondary protective barrier

___ Skin dose

___ Thermoluminescent dosimeters (TLD)

___ Time of occupancy (T)

___ Uncoltrolled area

___ Use (U)

___ Useful beam

___ Workload (W)

electrode is stationary, the other is moving. Exposure to ionization is determined by measurement of movement of the moving electrode.

j. The intensity of radiation as a given distance is inversely proportional to the square of the distance of the object from the source.

k. A measure of the radiation to the patient at the site of the patient's gonads.

l. A type of dosimeter consisting of radiation dosimetry film to determine the amount of exposure personnel have received.

m. A measure of the radiation to the patient's skin at the skin entrance surface.

n. The lowest dose of radiation that will maintain health with no ill effects.

o. Measurement of ionizing radiation doses to personnel.

p. A device that presets the on time for the tube to only five minute increments.

q. A measure of the radiation that has been absorbed into the patient's bone marrow.

r. As low as reasonably achievable.

s. Those states that have agreements with the Nuclear Regulatory Commission to take responsibility to enforce radiation protection guidelines through the states department of health.

t. An area occupied by radiation personnel.

u. The response of human cells exposed to radiation within minutes, days or weeks of exposure.

v. The lead-lined metal covering of the x-ray beam that serves to reduce leakage radiation.

w. Amount of radiation exposure.

x. Exposure control switch that is either a foot pedal or a hand switch.

y. The response of human cells exposed to radiation months to years after exposure.

z. Device that uses an ionization chamber to determine the level of exposure.

aa. Devices used to measure radiation doses.

(continued)

bb. The percentage of time in which the x-ray beam is energized and directed toward a particular wall.

cc. A measure of the radiation at the skin entrance surface.

dd. An area occupied by non-radiation personnel.

ee. A device containing lithium fluoride or calcium fluoride crystals to calculate the amount of personnel exposure.

ff. The accreditating body for radiographic facilities.

gg. The procedure of periodically activating the fluoroscopic tube.

REVIEW

1. List three early effects of radiation.

____

____

____

2. Name two precautions to be taken regarding the use of film badges.

____

____

____

3. List two advantages that TLDs have over film badges.

____

____

____

4. What primary advantage does a pocket dosimeter have over film badges and TLDs?

____

____

____

5. Guidelines for an institution's radiation monitoring program are established by:

a. hospital adminstrator

b. radiation safety officer

c. chief technologist

d. radiologist

6. How are present-day radiation protection guidelines established?

____

____

____

7. What is the effective dose-equivalent limit for a radiographer during any 13-week period?

a. 0.1 rem

b. 1 rem

c. 3 rem

d. 5 rem

8. A pregnant radiographer must be removed from all her duties.

a. True

b. False

9. List the formula which is used to calculate cumulative whole-body dose-equivalent limits.

____

____

____

10. How do gonadal doses for a PA chest and a lumbar spine compare?

____

____

____

11. Diagnostic radiology personnel may receive an annual effective dose-equivalent limit of ______________ for whole-body occupational exposure during routine operations.
   a. 0.5 rem (5 mSv)
   b. 1 rem (10 mSv)
   c. 5 rem (50 mSv)
   d. 10 rem (150 mSv)

12. Primary protective barriers must consist of _____ inch lead.
   a. 1/16
   b. 1/8
   c. 1/4
   d. 1/32

13. Shielding for an uncontrolled area maintains the exposure remains below _______ mR/week.
   a. 10
   b. 50
   c. 100
   d. 1,000

14. The exposure switch cord on a mobile unit must be long enough to allow the radiographer to stand back at least _______ feet during the exposure.
   a. 1
   b. 3
   c. 6
   d. 10

15. A Bucky slot cover should be at least ______________.
   a. 0.025 mm Pb equivalent
   b. 0.25 mm Pb equivalent
   c. 0.025 mm Al equivalent
   d. 0.25 mm Al equivalent

16. Protective lead gloves must have a minimum lead equivalency of at least ______ mm lead.
   a. 0.025
   b. 0.25
   c. 0.5
   d. 1.0

17. Which of the following people should be asked to hold a patient who needs physical support during the exposure?
   a. radiographer
   b. friend
   c. nurse
   d. doctor

EXPLORING THE WEB

1. Search the web for regulations regarding occupational exposure to radiation. What agencies regulate occupational exposure? Is your state an agreement state? Does your state enforce any regulations that differ from the national standard?

2. Search the web for additional information on safety in the workplace for radiographic technicians. What did you find? Did you discover any additional tips on maintaining a safe work environment?

3. Search the web for dosimetry. Explain how this works. Go to the web sites of companies that manufacture dosimeters. What products are available for the protection of the radiographic technician? What are the pros and cons of the products available?

CHAPTER 7

Protection of Patients

OBJECTIVES

Upon completion of this chapter, the reader should be able to:

- Discuss the importance of immobilization of the patient during an x-ray exposure
- Describe beam limiting apparatus and name the mechanism which best limits the x-ray beam
- Explain the purpose of x-ray beam filtration in diagnostic radiography
- State the reasons for using gonadal shielding during radiologic examinations and recognize the varieties of shields employed
- Discuss the necessity for using correct exposure factors for all radiologic procedures
- Demonstrate how the use of high speed film-screen combinations decreases radiographic exposure to the patient
- Explain the rationale for decreasing the number of repeat radiographs
- Discuss how patient exposure may be reduced during fluoroscopic procedures

LECTURE OUTLINE

Immobilization

____

____

____

Beam Restriction

____

____

____

(continued)

Kilovoltage

Irradiated material

Beam-Limiting Devices

Aperture Diaphragms

Cones

Collimators

X-Ray Beam Filtration

Gonadal Shielding

Flat Contact Shields

Shadow Shields

Shaped Contact Shields

Exposure and Technique Factors

(continued)

Kilovoltage Range

Milliamperage and Time

Film-Screen Considerations

Radiographic Film

Intensifying Screens

Patient Positioning

Grids

The Pregnant Patient

Repeat Radiographs

Fluoroscopic Procedures

Image Intensification Fluoroscopy

REVIEW OF KEY TERMS

Match the definition in the right column with the correct term from the left column.

___ Added filtration

___ Aperture diaphragm

___ Beam limitation device

___ Cones

___ Cumulative timing device

___ Flat contact shield

___ Gonad shielding

___ Half-value layer (HVL)

___ Image intensification flouroscopy

___ Inherent filtration

___ Intensifying screen

___ Intermittent flouroscopy

___ Involuntary motion

___ Shadow shield

___ Shaped contact shield

___ Source-image receptor distance (SID)

___ Total filtration

___ Variable-aperture collimator

___ Voluntary motion

a. A radiopaque material that is suspended from the tube housing that casts a shadow over the area that should not be exposed.

b. Lead devices that are shaped to fit over anatomical areas of the body to protect them from radiation exposure.

c. The area from the point of contact on the patient to the tube.

d. The maximum amount of limitation of the x-ray beam through the use of both added and inherent filtration methods.

e. A pair of upper and lower level lead shutters at right angles to each other that can be adjusted to limit the size field the x-ray beam will expose.

f. Movement controlled by the patient.

g. Rubberized lead strips that are placed over areas of the body to protect them from exposure.

h. Protection of the reproductive organs from radiation exposure.

i. The thickness of the material which will reduce the x-ray intensity to half of its original value.

j. A procedure used to increase the brightness of an image with a decreased exposure time.

k. Limiting the x-ray beam by use of a glass window in the tube and cooling oil surrounding the tube housing.

l. A device that serves to accelerate the action of x-rays by converting x-ray energy into visible light.

m. The procedure of periodically activating the flouro tube rather the having the tube continuously activated, thus reducing patient exposure and prolonging the life of the tube.

n. Movement of the patient that is beyond the patient's control.

o. Limiting the x-ray beam by the use of a piece of aluminum or aluminum equivalent outside the glass window of the housing tube.

p. A piece of flat lead with a hole in the center that attaches to the x-ray tube to confine the area of the beam.

q. Equipment that reduces the beam to only those areas in need of exposure.

r. A circular metal structure attached to the x-ray tube housing to restrict the x-ray beam to a predetermined size.

s. A device that presets the on time for the tube to only five minute increments.

REVIEW

1. In order to reduce the possibility of voluntary patient motion, the radiographer should:

a. use an immobilization device

b. reduce radiation exposure time

c. disregard any communication with patient

2. Which of the following is the most versatile type of x-ray beam limitation device?

a. aperture diaphragm

b. cone

c. cylinder

d. collimator

3. The function of filtration in diagnostic radiology is to:

a. decrease short wavelength radiation, thus increasing patient skin dose

b. increase beam hardness, thus increasing patient skin dose

c. increase beam hardness, thus reducing patient skin dose

d. decrease beam hardness, thus reducing patient skin dose

4. The most efficient type of male gonadal shielding for use during fluoroscopy is

a. shadow

b. flat contact

c. shaped contact

5. Which of the following combinations would reduce patient radiation dose during an x-ray examination?

a. higher kVp, lower mAs, increased filtration

b. lower kVp, higher mAs, decreased filtration

c. higher kVp, higher mAs, decreased filtration

d. lower kVp, lower mAs, increased filtration

6. Patient dose decreases when:

a. high-speed radiographic film is used in combination with high-speed intensifying screens

b. rare-earth intensifying screens are not used

c. low kVp techniques are used

d. non-screen film is used

7. Repeat radiographs result in:

a. no additional exposure to the patient

b. a double exposure of radiation to the patient

8. When a fluoroscopic image is amplified by an image intensifier, the resultant benefit is:

a. increased image brightness

b. radiologist must undergo dark adaptation

c. increased patient dose

d. decreased image brightness

9. The source-tabletop distance must not be less than ______________ for fixed fluoroscopes and not less than ______________ for mobile fluoroscopes.

a. 12 inches (30 cm), 6 inches (15 cm)

b. 15 inches (38 cm), 12 inches (30 cm)

c. 15 inches (38 cm), 9 inches (23 cm)

d. 18 inches (45 cm), 15 inches (38 cm)

EXPLORING THE WEB

1. Search the web for additional information on patient safety during radiographic procedures. Can you find any information to hand out to patients? Are you able to find any additional tips to ensure the safety of your patients?

2. Search the web for manufacturers of shielding equipment for use during radiographic procedures. List the pros and cons of the various types of equipment available. Are there any new technologies available to aid in patient protection and safety?

3. Search the web for tips and techniques in working with children who are undergoing radiographic procedures. Are there any special circumstances to be aware of with this population? How can you gain a child's trust? How can you help the child understand what the machinery will do?

SECTION III SELF-ASSESSMENT QUESTIONS

1. The effective dose-equivalent limit to the lens of the eye for a radiographer is:
   a. 1 rem
   b. 5 rem
   c. 10 rem
   d. 15 rem

2. Student radiographers under the age of 18 have an effective dose-equivalent limit of:
   a. 0.1 rem
   b. 0.5 rem
   c. 1 rem
   d. 5 rem

3. The total effective dose-equivalent limit to the fetus of a pregnant radiographer is:
   a. 0.1 rem
   b. 0.5 rem
   c. 1 rem
   d. 5 rem

4. Secondary protective barriers must have a lead thickness of:
   a. 1/64 inch
   b. 1/32 inch
   c. 1/16 inch
   d. 1/4 inch

5. How tall must primary protective barriers be?
   a. 3 feet
   b. 4 feet
   c. 7 feet
   d. 10 feet

6. Leakage radiation from the x-ray tube housing shall not exceed __________ at a distance of 3 feet from the tube.
   a. 1 mR/hr
   b. 10 mR/hr
   c. 100 mR/hr
   d. 1,000 mR/hr

7. The intensity of scattered radiation at 1 meter from the patient as compared to the intensity at the patient is:
   a. 0.001%
   b. 0.01%
   c. 0.1%
   d. 1%

8. The ______________ is the percentage of time which the x-ray beam is energized and directed toward a particular wall.
   a. occupancy factor (T)
   b. use factor (U)
   c. workload (W)

9. In accordance with NCRP Report #102, the lead equivalency of a lead apron must be:
   a. 0.12 mm
   b. 0.25 mm
   c. 0.5 mm
   d. 1 mm

10. In order to obey radiation safety regulations, what must the fluoroscopic exposure switch do?

a. abort fluoroscopic exposure after 5 minutes

b. be at the end of a six foot long expandable cord

c. make an audible sound during fluoroscopic exposure

d. be of the deadman type

11. In mobile fluoroscopy, the source-to-tabletop distance must not be less than ________.

a. 6 inches

b. 12 inches

c. 15 inches

d. 20 inches

12. A protective lead curtain that is at least __________ Pb equivalent must be positioned between the patient and fluoroscopy operator.

a. 0.025 mm

b. 0.25 mm

c. 2.5 mm

d. 25 mm

13. Fluoroscopic x-ray intensity at tabletop must not exceed:

a. 1 R/min

b. 5 R/min

c. 10 R/min

d. 100 R/min

14. If the exposure rate to a radiographer positioned 3 feet from a radiation source is 8 mR/min, what will be the dose to the radiographer at a distance of 6 feet from the source?

a. 1 mR/min

b. 2 mR/min

c. 16 mR/min

d. 32 mR/min

15. The exposure rate to a radiographer 3 ft from a radiation source is 40 R/hr. What distance from the source is necessary to decrease the exposure to 10 R/hr?

a. 1 ft

b. 2 ft

c. 6 ft

d. 12 ft

16. Radiation monitoring of personnel is required when personnel receive ____% of the annual effective dose-equivalent limit.

a. 0.01

b. 1

c. 5

d. 10

17. How many rems whole-body cumulative exposure would a 45-year-old radiographer be allowed to receive?

a. 15

b. 45

b. 35

d. 25

18. By what factor does a thyroid shield reduce radiation dose?

a. 10

b. 7

b. 4

d. 1

19. If an original intensity was 300 mR, and mAs is increased from 20 to 40 mAs, what would be the new intensity?

a. 150 mAs

b. 300 mAs

b. 600 mAs

d. 900 mAs

20. If 70 kVp produces an intensity of 200 mR, what will be the intensity at 100 kVp if no other factors are changed?

a. 100

b. 300

b. 408

d. 808

21. Voluntary patient motion is reduced by using shorter exposure times.

a. True

b. False

22. To within what percent of the SID must the collimator light and actual irradiated area be accurate?

a. 2%

b. 5%

c. 10%

d. 22%

23. The greatest beam limitation is accomplished when the cone/cylinder is ______________ and the diameter opening is ____________.

a. shorter, bigger

b. shorter, smaller

c. longer, bigger

d. longer, smaller

24. The function of a filter is to remove which of the following from the x-ray beam?

a. short wavelength radiation

b. long wavelength radiation

c. secondary radiation

d. scattered radiation

25. How much total filtration is required when using over 70 kVp?

a. 1.5 mm Al

b. 2.5 mm Al

c. 3.0 mm Al

d. 4.0 mm Al

26. Which of the following is the best type of gonadal shielding to use during a sterile field procedure?

a. flat contact shields

b. shadow shields

c. shaped contact shields

d. lead sheet

27. Which of the following exposure techniques will provide the least amount of radiation exposure to the patient?

a. 50 mAs, 90 kVp

b. 100 mAs, 90 kVp

c. 200 mAs, 50 kVp

d. 400 mAs, 50 kVp

28. Which of the following will best reduce radiation exposure to the patient?

a. use non-screen film

b. use rare-earth screens

c. decrease kVp

d. increase mAs

29. The number of repeat radiographs can be reduced by:

a. disregarding communication between patient and radiographer

b. eliminating voluntary patient motion using short exposure times

c. eliminating involuntary patient motion using immobilization devices

d. eliminating voluntary patient motion using immobilization devices

30. The tabletop exposure rate during fluoroscopy shall not exceed

a. 1 R/min

b. 5 R/min

c. 7 R/min

d. 10 R/min

31. A fluoroscope must be equipped with a cumulative timing device which times the radiation exposure and sounds an audible alarm after the fluoroscope has been energized for

a. 1 minute

b. 2 minutes

c. 3 minutes

d. 5 minutes

32. Which of the following is *not* a type of beam limitation device?

a. collimator

b. cone

c. filter

d. aperture diaphragm

33. The focal spot-to-table distance in fixed fluoroscopy must be

a. a minimum of 7 inches

b. a minimum of 12 inches

c. a minimum of 15 inches

d. a maximum of 12 inches

34. Half-value layer (HVL) is defined as the thickness of a designated absorber required to:

a. decrease the intensity of the primary beam by 50% of its initial value

b. decrease the intensity of the primary beam by 25% of its initial value

c. increase the intensity of the primary beam by 50% of its initial value

d. increase the intensity of the primary beam by 25% of its initial value

35. The minimum source-to-tabletop distance permitted in mobile fluoroscopy is

a. 4 inches

b. 9 inches

c. 12 inches

d. 15 inches

Appendix A: Answer Key

Chapter 1

Key Terms Review

c Direct effect
e Epilation
g Erythema
d Fractionation
f Indirect effect
j Law of Bergonie and Tribondeau
l Mutagenesis
i Rad
k Radioactivity
b Radiobiology
a Rem
m Reproductive failure
h Roengen (r)

Review

1. The Law of Bergonie and Tribondeau states:
   a. Correct. Stem cells are more radiosensitive than mature cells.
   b. Incorrect. Younger tissues and organs are more radiosensitive than older tissues and organs.
   c. Incorrect. The higher the metabolic activity of a cell, the more radiosensitive it is.
   d. Incorrect. The greater the proliferation and growth rate for tissues, the greater the radiosensitivity.

2. The _______ is the unit of dose equivalent or occupational exposure.
   a. Incorrect. The roentgen (R) is a measure of the ionization of air that is created by x- and gamma-radiation below 3 MeV.
   b. Correct. The rem (acronym for radiation equivalent man) represents the amount of radiation received by personnel.
   c. Incorrect. The rad (an acronym for radiation absorbed dose) describes the energy which is absorbed in matter from any type of ionizing radiation, and is considered the unit of absorbed dose.

3. Which of the following would be considered most radiosensitive?
   a. the fetus

4. The unit of radiation quantity is the ___.
   c. roentgen

5. Which researcher discovered mutations related to exposure to ionizing radiation?
   a. Muller

Chapter 2

Key Terms Review

c Amino acid
j Anabolism
v Anaphase
d Autosome
w Catabolism
k Centromere
x Chromatid
y Chromatin
u Chromosome
z Deoxyribonucleic acid (DNA)
l Diploid
aa Duplication

bb Enzyme
e G_1
f G_2
cc Gamete
m Gene
b Haploid
t Interphase
dd Macromolecule
ee Meiosis
a Messenger RNA
g Metaphase
s Mitosis
n Nuclear membrane
o Nucleolus
p Organism
h Polymer
gg Prophase
hh Protoplasm
ii Ribosomal RNA
r Ribonucleic acid (RNA)
i S-phase
q Telophase
ff Transfer RNA

Review

1. A combination of two or more tissues which are combined to perform a specific function is a definition for:

a. Incorrect. The cell is the basic unit of structure and function of all living things.

b. Incorrect. Tissues are collections of similar cells that work together in performing a particular function.

c. Correct. An organ is an combination of two or more tissues which are combined to perform a specific function.

d. Incorrect. Organs that have similar or related functions compose a body system.

2. ______________ assist in growth, construct new tissues, and repair injured or worn-out cells.

a. Correct. Proteins assist in growth, construct new tissues, and repair injured or worn-out cells.

b. Incorrect. Enzymes act as organic catalysts. They control the numerous chemical reactions that occur in cells.

c. Incorrect. Lipids function in storing energy, insulate our bodies from cold, and assist with the digestive process.

d. Incorrect. Carbohydrates are the major source of cell energy.

3. The location of genetic information is in the:

a. Incorrect. A cell membrane forms the outer confines of the cell. It functions to separate the cell's cytoplasm from its exterior surroundings and also from adjacent cells.

b. Incorrect. Endoplasmic reticulum (ER) assists in channeling proteins and lipids into and out of the nucleus.

c. Incorrect. Mitochondria supply the cell's energy. They are nicknamed the "powerhouses" of the cell.

d. Correct. Chromosomes are contained within the nucleus. Chromosomes contain the human hereditary blueprint—DNA.

4. Which of the following is the RNA nucleotide base that pairs with adenine in DNA synthesis?

a. Incorrect. RNA is a single-stranded nucleotide, with uracil pairing with adenine in place of thymine.

b. Correct. RNA is a single-stranded nucleotide, with uracil pairing with adenine in place of thymine.

c. Incorrect. RNA is a single-stranded nucleotide, with uracil pairing with adenine in place of thymine.

d. Incorrect. RNA is a single-stranded nucleotide, with uracil pairing with adenine in place of thymine.

5. The normal diploid or 2n number in humans is:

a. Incorrect. The normal number of chromosomes for humans is 46. This number is termed the diploid number, or 2n. The normal haploid or n number is 23.

b. Incorrect. The normal number of chromosomes for humans is 46. This number is termed the diploid number, or 2n. The normal haploid or n number is 23.

c. Correct. The normal number of chromosomes for humans is 46. This number is termed the diploid number, or 2n. The normal haploid or n number is 23.

d. Incorrect. The normal number of chromosomes for humans is 46. This number is termed the diploid number, or 2n. The normal haploid or n number is 23.

6. In which phase of mitosis do the chromosomes line up at the equator of the cell?

a. Incorrect. During prophase, the centrioles migrate toward opposite poles of the cell, producing spindle fibers which extend across the cell's equator.

b. Correct. During metaphase, the paired chromosomes are lined up at the equator of the cell.

c. Incorrect. During anaphase, the centromeres divide, and the sister chromatids detach as they are pulled to an opposite pole.

d. Incorrect. During telophase, the sets of chromosomes become much longer, thinner, and indistinct as they reach the poles of the cell. The DNA unravels to form chromatin.

7. During meiosis, or reduction division,

a. Correct. During meiosis, the cell divides twice in succession, but chromosomes are duplicated only one time.

b. Incorrect. During meiosis, the cell divides twice in succession, but chromosomes are duplicated only one time.

c. Incorrect. During meiosis, the cell divides twice in succession, but chromosomes are duplicated only one time.

d. Incorrect. During meiosis, the cell divides twice in succession, but chromosomes are duplicated only one time.

Chapter 3

Key Terms Review

b Deletion
e Dicentric
l Dose-response relationship (curve)
j Effectiveness
s Free radical
u HeLa
m Interphase death
t Linear dose-response curve
v Linear quadratic dose-response curve
f Linear energy transfer (LET)
g Mutation
r Nonthreshold
q Oxygen effect
k Point mutation
h Radiolysis
n Radiosensitivity
p Relative biolgic effectiveness (RBE)
o Ring
c Sigmoid
i System
j Threshold
d Translocation

Review

1. Which of the following cell groups are considered highly radiosensitive?

a. Correct. Cells which are considered highly radiosensitive include: lymphocytes, spermatogonia, erythroblasts, and intestinal crypt cells.

b. Incorrect. Endothelial cells, osteoblasts, spermatids, and fibroblasts have an intermediate radiosensitivity.

c. Incorrect. Muscle and nerve cells, and chondrocytes have low radiosensitivity.

2. Relative biologic effectiveness (RBE):

a. Incorrect. A measure of the rate at which energy is deposited as a charged particle travels through matter describes Linear Energy Transfer (LET).

b. Correct. Relative biologic effectiveness (RBE) is a comparison of a dose of test radiation to a dose of 250-keV x-ray that produces the same biologic response.

c. Incorrect. The dose of radiation that produces a given biologic response under anoxic conditions divided by the dose of radiation that produces the same biologic response under aerobic conditions describes Oxygen Enhancement Ratio (OER).

3. Which of the following body molecules are most commonly acted upon directly by ionizing radiation to produce indirect effects?

a. Incorrect. Although DNA is the most critical target of radiation, it is the irradiating of water, which causes indirect effects, that is the principal cause of effects from radiation.

b. Incorrect. Although DNA is the most critical target of radiation, it is the irradiating of water, which causes indirect effects, that is the principal cause of effects from radiation.

c. Incorrect. Although DNA is the most critical target of radiation, it is the irradiating of water, which causes indirect effects, that is the principal cause of effects from radiation.

d. Correct. Although DNA is the most critical target of radiation, it is the irradiating of water, which causes indirect effects, that is the principal cause of effects from radiation.

4. The types of DNA molecule damage include:

a, b, c, d, and e are all correct.

5. At low doses of radiation, most cellular radiation damage is the result of:

a. Incorrect. At low doses of radiation, point lesions are regarded to be the cellular radiation damage that are responsible for the late radiation effects which are observed at the whole-body level.

b. Incorrect. At low doses of radiation, point lesions are regarded to be the cellular radiation damage that are responsible for the late radiation effects which are observed at the whole-body level.

c. Correct. At low doses of radiation, point lesions are regarded to be the cellular radiation damage that are responsible for the late radiation effects which are observed at the whole-body level.

6. Nonthreshold:

a. Incorrect. That an observed response is directly proportional to the dose is a description of linear.

b. Incorrect. That an observed response is not directly proportional to the dose is a description of nonlinear.

c. Incorrect. That there is a radiation level reached under which there would be no effects observed is a description of threshold.

d. Correct. Nonthreshold assumes that any radiation dose produces an effect.

7. According to target theory:

a. Incorrect. According to the target theory, there will be cell death only if the cell's target molecule is inactivated. It is theorized that DNA is the critical molecular target. Both direct and indirect effects cause hits.

b. Incorrect. According to the target theory, there will be cell death only if the cell's target molecule is inactivated. It is theorized that DNA is the critical molecular target. Both direct and indirect effects cause hits.

c. Incorrect. According to the target theory, there will be cell death only if the cell's target molecule is inactivated. It is theorized that DNA is the critical molecular target. Both direct and indirect effects cause hits.

d. Correct. According to the target theory, there will be cell death only if the cell's target molecule is inactivated. It is theorized that DNA is the critical molecular target. Both direct and indirect effects cause hits.

8. Which portion of the cell survival curve proves that some damage must accumulate before cell death can occur?

a. Correct. The shoulder of the cell survival curve shows that some damage must accrue before there can be cell death. The accumulated damage is called sublethal damage. The wider the shoulder, the more sublethal damage the cell can endure.

b. Incorrect. The shoulder of the cell survival curve shows that some damage must accrue before there can be cell death. The accumulated damage is called sublethal damage. The wider the shoulder, the more sublethal damage the cell can endure.

c. Incorrect. The shoulder of the cell survival curve shows that some damage must accrue before there can be cell death. The accumulated damage is called sublethal damage. The wider the shoulder, the more sublethal damage the cell can endure.

SECTION I SELF-ASSESSMENT ANSWERS

1. a. Correct. Mitosis is the term for cell division of somatic cells.

b. Incorrect. Mitosis is the term for cell division of somatic cells.

c. Incorrect. Mitosis is the term for cell division of somatic cells.

d. Incorrect. Mitosis is the term for cell division of somatic cells.

2. a. Incorrect. Metaphase is considered the most radiosensitive stage of mitosis.

b. Incorrect. Metaphase is considered the most radiosensitive stage of mitosis.

c. Correct. Metaphase is considered the most radiosensitive stage of mitosis.

d. Incorrect. Metaphase is considered the most radiosensitive stage of mitosis.

3. a. Incorrect. The DNA molecule is described as being in the shape of a double helix.

b. Incorrect. The DNA molecule is described as being in the shape of a double helix.

c. Incorrect. The DNA molecule is described as being in the shape of a double helix.

d. Correct. The DNA molecule is described as being in the shape of a double helix.

4. a. Incorrect. The normal human cell contains 23 matched pairs of chromosomes, or 46 total.
   b. Correct. The normal human cell contains 23 matched pairs of chromosomes, or 46 total.
   c. Incorrect. The normal human cell contains 23 matched pairs of chromosomes, or 46 total.
   d. Incorrect. The normal human cell contains 23 matched pairs of chromosomes, or 46 total.

5. a. Incorrect. Interphase is considered the "resting stage" of cell division.
   b. Incorrect. Interphase is considered the "resting stage" of cell division.
   c. Incorrect. Interphase is considered the "resting stage" of cell division.
   d. Correct. Interphase is considered the "resting stage" of cell division.

6. a. Correct. Altering a base sequence of DNA would result in a gene mutation.
   b. Incorrect. Altering a base sequence of DNA would result in a gene mutation.
   c. Incorrect. Altering a base sequence of DNA would result in a gene mutation.
   d. Incorrect. Altering a base sequence of DNA would result in a gene mutation.

7. a. Incorrect. Amino acids are considered the "building blocks" of proteins.
   b. Incorrect. Amino acids are considered the "building blocks" of proteins.
   c. Correct. Amino acids are considered the "building blocks" of proteins.
   d. Incorrect. Amino acids are considered the "building blocks"of proteins.

8. a. Incorrect. The majority of RNA is found in the cytoplasm.
   b. Incorrect. The majority of RNA is found in the cytoplasm.
   c. Incorrect. The majority of RNA is found in the cytoplasm.
   d. Correct. The majority of RNA is found in the cytoplasm.

9. a. Incorrect. Genes are the areas of the DNA molecule that determine cell characteristics.
   b. Incorrect. Genes are the areas of the DNA molecule that determine cell characteristics.
   c. Correct. Genes are the areas of the DNA molecule that determine cell characteristics.
   d. Incorrect. Genes are the areas of the DNA molecule that determine cell characteristics.

10. a. Correct. Oxygen retention increases radiosensitivity.
   b. Incorrect. Oxygen retention increases radiosensitivity.
   c. Incorrect. Oxygen retention increases radiosensitivity.
   d. Incorrect. Oxygen retention increases radiosensitivity.

11. a. Incorrect. A free radical has one or more unpaired electrons and is usually chemically reactive.
   b. Incorrect. A free radical has one or more unpaired electrons and is usually chemically reactive.
   c. Correct. A free radical has one or more unpaired electrons and is usually chemically reactive.
   d. Incorrect. A free radical has one or more unpaired electrons and is usually chemically reactive.

12. a. Incorrect. Genes contain the unit of heredity.
   b. Incorrect. Genes contain the unit of heredity.
   c. Correct. Genes contain the unit of heredity.
   d. Incorrect. Genes contain the unit of heredity.

13. a. Incorrect. After chromosomes split in half, the two halves are called chromatids.
   b. Correct. After chromosomes split in half, the two halves are called chromatids.
   c. Incorrect. After chromosomes split in half, the two halves are called chromatids.
   d. Incorrect. After chromosomes split in half, the two halves are called chromatids.

14. a. Incorrect. In order for protein synthesis to occur, m-RNA carries information to the ribosome.
   b. Incorrect. In order for protein synthesis to occur, m-RNA carries information to the ribosome.
   c. Incorrect. In order for protein synthesis to occur, m-RNA carries information to the ribosome.
   d. Correct. In order for protein synthesis to occur, m-RNA carries information to the ribosome.

15. a. Incorrect. T-RNA carries amino acids in order to synthesize proteins.
   b. Incorrect. T-RNA carries amino acids in order to synthesize proteins.
   c. Incorrect. T-RNA carries amino acids in order to synthesize proteins.
   d. Correct. T-RNA carries amino acids in order to synthesize proteins.

16. a. Correct. The majority of a cell's genetic information is found in the nucleus.

b. Incorrect. The majority of a cell's genetic information is found in the nucleus.

c. Incorrect. The majority of a cell's genetic information is found in the nucleus.

d. Incorrect. The majority of a cell's genetic information is found in the nucleus.

17. a. Incorrect. Cell division of germ cells is named meiosis.

b. Correct. Cell division of germ cells is named meiosis.

c. Incorrect. Cell division of germ cells is named meiosis.

d. Incorrect. Cell division of germ cells is named meiosis.

18. a. Incorrect. The haploid number refers to the single set of chromosomes in a germ cell.

b. Correct. The haploid number refers to the single set of chromosomes in a germ cell.

c. Incorrect. The haploid number refers to the single set of chromosomes in a germ cell.

d. Incorrect. The haploid number refers to the single set of chromosomes in a germ cell.

19. a. Incorrect. Linear energy transfer measures the rate of energy lost along the track of an ionizing particle.

b. Correct. Linear energy transfer measures the rate of energy lost along the track of an ionizing particle.

c. Incorrect. Linear energy transfer measures the rate of energy lost along the track of an ionizing particle.

d. Incorrect. Linear energy transfer measures the rate of energy lost along the track of an ionizing particle.

20. a. Incorrect. Interphase death means that cells die before they leave the interphase portion of cell division.

b. Correct. Interphase death means that cells die before they leave the interphase portion of cell division.

c. Incorrect. Interphase death means that cells die before they leave the interphase portion of cell division.

d. Incorrect. Interphase death means that cells die before they leave the interphase portion of cell division.

21. a. Correct. Sublethal damage describes cell damage from radiation that is not sufficient to kill the cell.

b. Incorrect. Sublethal damage describes cell damage from radiation that is not sufficient to kill the cell.

c. Incorrect. Sublethal damage describes cell damage from radiation that is not sufficient to kill the cell.

d. Incorrect. Sublethal damage describes cell damage from radiation that is not sufficient to kill the cell.

22. a. Incorrect. Free radicals are a threat to humans since they have been observed to produce toxic effects.

b. Incorrect. Free radicals are a threat to humans since they have been observed to produce toxic effects.

c. Incorrect. Free radicals are a threat to humans since they have been observed to produce toxic effects.

d. Correct. Free radicals are a threat to humans since they have been observed to produce toxic effects.

23. a. Incorrect. According to the target theory, DNA is thought to be the principal target of cell damage.

b. Correct. According to the target theory, DNA is thought to be the principal target of cell damage.

c. Incorrect. According to the target theory, DNA is thought to be the principal target of cell damage.

d. Incorrect. According to the target theory, DNA is thought to be the principal target of cell damage.

24. a. Incorrect. Indirect effect refers to ionizing events in one location which can produce effects at a distant location.

b. Incorrect. Indirect effect refers to ionizing events in one location which can produce effects at a distant location.

c. Correct. Indirect effect refers to ionizing events in one location which can produce effects at a distant location.

d. Incorrect. Indirect effect refers to ionizing events in one location which can produce effects at a distant location.

25. a. Correct. The fetus would be considered the most radiosensitive, as it would contain the most amount of immature cells.
   b. Incorrect. The fetus would be considered the most radiosensitive, as it would contain the most amount of immature cells.
   c. Incorrect. The fetus would be considered the most radiosensitive, as it would contain the most amount of immature cells.
   d. Incorrect. The fetus would be considered the most radiosensitive, as it would contain the most amount of immature cells.

26. a. Incorrect. Radiolysis is the term used to describe the separation of water into hydrogen and oxygen following irradiation.
   b. Incorrect. Radiolysis is the term used to describe the separation of water into hydrogen and oxygen following irradiation.
   c. Incorrect. Radiolysis is the term used to describe the separation of water into hydrogen and oxygen following irradiation.
   d. Correct. Radiolysis is the term used to describe the separation of water into hydrogen and oxygen following irradiation.

27. a. Correct. The rem (acronym for radiation equivalent man) is the unit of measure used for expressing occupational exposure.
   b. Incorrect. The rem (acronym for radiation equivalent man) is the unit of measure used for expressing occupational exposure.
   c. Incorrect. The rem (acronym for radiation equivalent man) is the unit of measure used for expressing occupational exposure.
   d. Incorrect. The rem (acronym for radiation equivalent man) is the unit of measure used for expressing occupational exposure.

28. a. Incorrect. Metaphase is considered the most radiosensitive stage of mitosis.
   b. Incorrect. Metaphase is considered the most radiosensitive stage of mitosis.
   c. Correct. Metaphase is considered the most radiosensitive stage of mitosis.
   d. Incorrect. Metaphase is considered the most radiosensitive stage of mitosis.

29. a. Correct. Increasing LET increases RBE.
   b. Incorrect. Increasing LET increases RBE.
   c. Incorrect. Increasing LET increases RBE.
   d. Incorrect. Increasing LET increases RBE.

30. a. Incorrect. In DNA, adenine pairs with thymine, and guanine pairs with cytosine.
   b. Correct. In DNA, adenine pairs with thymine, and guanine pairs with cytosine.
   c. Incorrect. In DNA, adenine pairs with thymine, and guanine pairs with cytosine.
   d. Incorrect. In DNA, adenine pairs with thymine, and guanine pairs with cytosine.

31. a. Incorrect. Base damage would be most likely to cause abnormalities in base sequences, and thus cell mutation.
   b. Incorrect. Base damage would be most likely to cause abnormalities in base sequences, and thus cell mutation.
   c. Incorrect. Base damage would be most likely to cause abnormalities in base sequences, and thus cell mutation.
   d. Correct. Base damage would be most likely to cause abnormalities in base sequences, and thus cell mutation.

Chapter 4

Key Terms Review

a Aberration
f Acentric
aa Alopecia
m Anemia
o Anomaly
n Atrophy
i Crypts of Lieberkuhn
p Cytopenia
bb Dermis
g Desquamation
cc Decentric
b Edema
dd Epidermis
q Erythroblasts
ee Follicles
t Granulocytes
v Hemopoietic
gg Inflammation
x Karyotype
h Latent

ii LD50/30
kk LD50/60
c Lesion
l Manifest
nn Maturation depletion
mm Megakaryocytes
ff Meningitis
hh Myelocytes
jj Necrosis
ll Oogonla
y Prodromal
z Radiation cataractogenesis
w Radioresistant
pp Sebaceous
k SED50
u Skin erythema dose (SED)
oo Somatic
qq Spermatogonia
j Stem cell
s Stroma
rr Subcutaneous
e Syndrome
r Threshold dose
d Vasculitis

Review

1. An organ must have been acutely exposed; the total body must be exposed; external rather than internal penetrating sources cause the radiation syndrome.

2. Life span shortening.

3. The approximate LD50/30 for humans is 250–450 rads.

4. The acute radiation syndromes are the hematologic/hematopoietic/bone marrow; gastrointestinal (GI); and central nervous system (CNS) syndromes. Dose ranges for the hematologic syndrome are 100–1000 R; ranges for the GI syndrome are 600–10,000 R; ranges for the CNS syndrome are over 10,000 R.

5. Prodromal stage, latent stage, manifest illness stage, and recovery/death.

6. Cause of death from the bone marrow syndrome is from destruction of the bone marrow which causes infection and hemorrhage, and ultimately, death.

7. Damage sustained to the GI tract and bone marrow causes infection, dehydration and electrolyte imbalance, which ultimately leads to death.

8. Damage to the blood vessels which supply the central nervous system cause edema in the cranial vault, vasculitis, and meningitis. It is theorized that death is from increased intra-cranial pressure as a result of elevated fluid content caused by the earlier mentioned changes.

9. The SED50 is the skin erythema dose necessary to affect 50% of people so irradiated, and is estimated to be approximately 600 R (6 Gy). Prior to the introduction of radiation units, the skin erythema dose (SED) was used as a measure of the amount of radiation a person had received.

10. Radiation-induced cataracts follow a threshold, nonlinear dose-response relationship. The threshold dose is thought to be approximately 200 rads (2 Gy).

11. The main effect radiation has on the male gonads is damage to and reduction in number of the spermatogonia, which ultimately leads to depletion of mature sperm, a process called maturation depletion.

12. In both the male and female, doses of 200 rads produce temporary sterility. An acute dose of approximately 500 to 600 rads will cause permanent sterility.

13. The male spermatogonia are continuously reproducing, whereas the female oogonia multiply only in the fetus.

14. Megakaryocytes are the least sensitive. Next in sensitivity are the myelocytes. The most sensitive stem cells are the erythroblasts.

15. If the neutrophils and lymphocytes are depleted, people are more vulnerable to infections. If platelets are depleted, there is an increased chance of hemorrhage. If red blood cells are depleted, the person will become anemic.

16. Chromosome aberrations occur *before* DNA synthesis. If chromosomal damage is not repaired before DNA synthesis, a chromosomal break will be replicated. If this is the case, both daughter cells will inherit a damaged chromatid.

17. Single-break chromosome aberrations include terminal deletions, inversions, and duplications. Double-break chromosome aberrations include interstitial deletions, inversions, duplications, and translocations.

18. Somatic mutations affect only that person who has the mutation. Genetic mutations may affect reproductive organs or the parent's gametes, which may affect future progeny.

19. Virtually all types of chromosome aberrations can be radiation-induced. The rate of aberrations is related to total radiation dose and dose rate, and these are thought to be nonthreshold. Aberrations can be produced by low and high radiation doses. Radiation-induced chromosome effects appear to be nonspecific, and are considered to be undesirable.

Chapter 5

Key Terms Review

p Absolute risk model
bb Acute
y Ankylosing spondylitis
o BEIR
u Benign
q Carcinogenic
r Carcinoma
gg Chronic
s Doubling dose
t Excess risk
aa Fibrosis
v Follicular
g Genetically significant dose (GSD)
h Hypoxic
i Leukemia
w Loci
cc Lymphoid
dd Malignant
b Myeloid
a Neoplasm
ee Neruoblasts
c Nonstochastic
x Osteosarcoma
ff Papillary
j Parenchymal
k Progeny
l Radiocarcinogenesis
m Radium
n Radon
f Relative risk model
e Stochastic
d Turner's syndrome
z zygote

Review

1. Atomic bomb survivors, medically exposed patients, occupationally exposed personnel and populations who receive high natural background exposure

2. The relative (multiplicative) risk model theorizes that as a person ages, his risk of developing cancer increases. The absolute (additive) risk model predicts a continual increase in risk which is not dependent on the spontaneous age specific cancer risk at the time of exposure.

3. Relative risk is the ratio of cancer incidence in an exposed population to that of an unexposed population.

The formula for relative risk $= \frac{\text{observed cases}}{\text{expected cases}}$.

Absolute risk is expressed in terms of number of cases/10^6 persons/rad/year. This risk model assumes a linear dose-response relationship. Excess risk describes the number of excess cases observed compared with the expected spontaneous occurrence. The formula for excess risk = observed cases - expected cases.

4. In Hiroshima, 61 leukemia deaths were observed as compared with the expected incidence of 12.

5. The incidence differences seen are caused by the types of radiation that each city was subjected to. Nearly 90% of the dose at Nagasaki was from x-rays, whereas at Hiroshima nearly half the dose was from x-rays and half was from neutrons. As neutrons have a higher RBE than x-rays, they are more biologically damaging than x-rays.

6. Radiation-induced leukemia appears to be linear, nonthreshold, has a latent period of 4–7 years, and an at-risk period of 15–20 years.

7. Radiation-induced skin cancer follows a threshold dose-response relationship, and has a latent period of 5–10 years. Radiation-induced skin carcinomas are not present in current radiology personnel.

8. Thyroid cancer is responsible for approximately 12% of the deaths attributed to radiation-induced malignancies. Females have approximately a 3–5 times greater risk for radiation-induced thyroid cancer than males because of hormonal influences on thyroid function. Dose-response relationships appear to be linear, nonthreshold.

9. Radiation-induced breast cancer follows a linear dose-response relationship. The latent period varies from 10–40 years. Absolute risk is estimated to be approximately six cases/10^6 persons/rad/year.

10. Data analysis of the painters indicated an overall relative risk of 122:1. The absolute risk is 0.11 cases/10^6 persons/rad/year. Osteosarcoma follows a linear quadratic dose-response relationship.

11. Dose-response relationships for radiation-induced lung cancer are linear nonthreshold. Based on data from over 4,000 uranium miners, an absolute risk of 1.3 cases/10^6 persons/rad/year has been calculated.

12. A study of radiologists provides the only evidence of human life span shortening. Although animal studies have been performed that suggest specific life span shortening could occur, human life span shortening appears to be nonspecific, that is, the cancer is caused by effects other than the radiation.

13. Muller's data indicated that: 1) there was not an increase in the quality of types of mutations; 2) the majority of mutations were recessive; 3) a threshold was not observed; 4) mutations were single-hit and cumulative. Russell's data differed in that a dose-rate effect was observed. A dose which was extended over a long period of time demonstrated fewer genetic effects than one large dose.

14. The GSD speculates that the long-term radiation effects can be averaged over an entire population. The GSD is calculated by using a formula that examines the average gonadal dose per examination, number of people receiving x-ray examinations, total number of people in the population, and anticipated number of children per person. The GSD is an average calculated from actual doses that are received by the entire population.

15. The doubling dose is that radiation dose which doubles the number of spontaneous mutations. The doubling dose for humans is estimated to be in the range of 50 to 250 rads (0.5 and 2.5 Gy) for each generation.

16. Even though some cancers are genetic in origin, it has not been proven that radiation has the capability to influence the inheritance of cancer susceptibility. Doses associated with diagnostic and occupational radiation exposures would not be anticipated to cause any significant risk to their offspring.

17. Total dose, dose rate, radiation quality, and stage of fetal development

18. The all-or-nothing response occurs during the pre-implantation stage. Either there is a spontaneous radiation-induced abortion, or no effects are observed.

19. Data obtained from animal experiments show an increase in the spontaneous abortion rate after doses of 5–10 rads (50–100 mGy) given during the pre-implantation stage. Following implantation, doses of at least 25 rads (250 mGy) are necessary to cause prenatal death. The natural occurrence of spontaneous abortion is approximately 25–50%.

20. The development of the CNS takes place over a much longer gestational period in humans than in animals and, therefore, is more likely to be a target for radiation-induced injury. The CNS is not fully developed until approximately age 12 years. All cases of human irradiation in utero which resulted in gross malformations have been accompanied by CNS aberrations and/or retarded growth.

21. The three stages of the human gestational period are pre-implantation stage, major organogenesis, and fetal growth stage.

22. Children irradiated in utero between the 8th and 25th week demonstrated lower than normal IQ scores. This decrease in IQ was dependent on a presumed threshold dose of below 25 rads. A high incidence of microcephaly was also observed.
Similar data were obtained from children whose mothers had received diagnostic or therapeutic irradiation during pregnancy. In a study of 28 children who were irradiated in utero from pelvic radium or x-ray therapy, 20 of these children were mentally retarded, 16 were also microcephalic. Other deformities which have been observed from irradiation in utero include abnormal extremities, hydrocephaly, spina bifida, blindness, cataracts, and microphthalmia.

23. There are thresholds for the majority of radiation-induced congenital abnormalities. Doses of less than 10 rad carry negligible risk. In current diagnostic procedures, fetal irradiation rarely exceeds 5 rad (50 mGy). This dose range has not demonstrated any significant risk for congenital malformation or retarded growth.

24. The current estimates for leukemia mortality are 2–3/10,000 after 1 rad (1 cGy) of low LET radiation. Solid tumors occur at approximately the same incidence. This brings the combined mortality from in utero irradiation to approximately 4–6/10,000 per rad (1 cGy). These mortality figures are thought to be identical to those cancers that occur spontaneously and in insufficient numbers to be easily identified in an exposed population.

25. The current effective dose-equivalent limit to the fetus is no more than 500 mrems (5 mSv) during the gestational period, provided that the dose rate does not exceed 50 mrems (0.5 mSv) in any one month.

26. Prior to diagnostic examinations that involve ionizing radiation, all fertile female patients should be questioned as to whether there is the possibility that they might be pregnant. If it is known that the patient is pregnant and alternative diagnostic procedures are not appropriate, the risks and benefits of the examination should be discussed with the patient prior to the examination.

27. Stochastic effects demonstrate an increase in effects in proportion to the radiation dose of the whole population. They are considered nonthreshold, and are associated with the linear and linear quadratic dose-response curves. Stochastic effects are regarded as the main health risk from low-dose radiation from exposures in the diagnostic radiology department. Examples of stochastic effects include radiation-induced cancer and radiation-induced genetic effects.

SECTION II
SELF-ASSESSMENT ANSWERS

1. a. Incorrect. The LD50/30 is closest to 300 rad.

b. Incorrect. The LD50/30 is closest to 300 rad.

c. Incorrect. The LD50/30 is closest to 300 rad.

d. Correct. The LD50/30 is closest to 300 rad.

2. a. Incorrect. Effects are dependent on LET.

b. Incorrect. For sterility the dose-response relationship is linear, threshold.

c. Incorrect. In the male the spermatogonia are more radiosensitive than the spermatocyte.

d. Correct. Radiation doses as low as 10 rad have caused reductions in the number of germ cells.

3. a. Incorrect. The LD50/30 is the dose required to kill half the people in 30 days.

b. Incorrect. The LD50/30 is the dose required to kill half the people in 30 days.

c. Correct. The LD50/30 is the dose required to kill half the people in 30 days.

d. Incorrect. The LD50/30 is the dose required to kill half the people in 30 days.

4. a. Incorrect. Genetic damage is considered a late response of radiation exposure.

b. Incorrect. Leukemia is considered a late response of radiation exposure.

c. Incorrect. Life span shortening is considered a late response of radiation exposure.

d. Correct. Cytogenetic damage is considered an early response of radiation exposure.

5. a. Incorrect. Human radiation lethality follows a nonlinear, threshold dose-response relationship.

b. Correct. Human radiation lethality follows a nonlinear, threshold dose-response relationship.

c. Incorrect. Human radiation lethality follows a nonlinear, threshold dose-response relationship.

d. Incorrect. Human radiation lethality follows a nonlinear, threshold dose-response relationship.

6. a. Incorrect. A dose of approximately 200 rad to the ovaries is necessary to produce permanent sterility.

b. Incorrect. A dose of approximately 200 rad to the ovaries is necessary to produce permanent sterility.

c. Incorrect. A dose of approximately 200 rad to the ovaries is necessary to produce permanent sterility.

d. Correct. A dose of approximately 200 rad to the ovaries is necessary to produce permanent sterility.

7. a. Incorrect. The GI syndrome has a mean survival time that is independent of dose.

b. Correct. The GI syndrome has a mean survival time that is independent of dose.

c. Incorrect. The GI syndrome has a mean survival time that is independent of dose.

8. a. Incorrect. The approximate skin erythema dose required to affect 50% of persons so irradiated is approximately 600 rad.

b. Incorrect. The approximate skin erythema dose required to affect 50% of persons so irradiated is approximately 600 rad.

c. Correct. The approximate skin erythema dose required to affect 50% of persons so irradiated is approximately 600 rad.

d. Incorrect. The approximate skin erythema dose required to affect 50% of persons so irradiated is approximately 600 rad.

9. a. Correct. Breast cancer is considered a late response to radiation exposure.

b. Incorrect. Skin erythema which occurs 2 weeks postexposure is considered an early response to radiation exposure.

c. Incorrect. Intestinal distress which occurs 1 week postexposure is considered an early response to radiation exposure.

d. Incorrect. Chromosome aberrations are considered an early response to radiation exposure.

10. a. Correct. Cataracts are not an acute local tissue effect.

b. Incorrect. Epilation is considered an acute local tissue effect.

c. Incorrect. Temporary sterility is considered an acute local tissue effect.

d. Incorrect. Skin erythema is considered an acute local tissue effect.

11. a. Incorrect. Death caused by a single dose of whole body irradiation involves damage primarily to the bone marrow.

b. Correct. Death caused by a single dose of whole body irradiation involves damage primarily to the bone marrow.

c. Incorrect. Death caused by a single dose of whole body irradiation involves damage primarily to the bone marrow.

d. Incorrect. Death caused by a single dose of whole body irradiation involves damage primarily to the bone marrow.

12. a. Incorrect. The acute radiation syndrome is the clinical pattern produced by a whole body dose of radiation which is given in a period of seconds to minutes.

b. Incorrect. The acute radiation syndrome is the clinical pattern produced by a whole body dose of radiation which is given in a period of seconds to minutes.

c. Correct. The acute radiation syndrome is the clinical pattern produced by a whole body dose of radiation which is given in a period of seconds to minutes.

d. Incorrect. The acute radiation syndrome is the clinical pattern produced by a whole body dose of radiation which is given in a period of seconds to minutes.

13. a. Incorrect. GSD stands for genetically significant dose, which is an average calculated from actual gonadal doses received by an entire population.

b. Incorrect. GSD stands for genetically significant dose, which is an average calculated from actual gonadal doses received by an entire population.

c. Incorrect. GSD stands for genetically significant dose, which is an average calculated from actual gonadal doses received by an entire population.

d. Correct. GSD stands for genetically significant dose, which is an average calculated from actual gonadal doses received by an entire population.

14. a. Incorrect. There is not a minimum level below which no genetic or somatic damage would occur.

b. Incorrect. There is not a minimum level below which no genetic or somatic damage would occur.

c. Incorrect. There is not a minimum level below which no genetic or somatic damage would occur.

d. Correct. There is not a minimum level below which no genetic or somatic damage would occur.

15. a. Correct. The greatest radiation hazard to a fetus is during the first trimester.

b. Incorrect. The greatest radiation hazard to a fetus is during the first trimester.

c. Incorrect. The greatest radiation hazard to a fetus is during the first trimester.

16. a. Incorrect. The principal response of blood caused by radiation exposure is a decrease in cell number.

b. Incorrect. The principal response of blood caused by radiation exposure is a decrease in cell number.

c. Correct. The principal response of blood caused by radiation exposure is a decrease in cell number.

d. Incorrect. The principal response of blood caused by radiation exposure is a decrease in cell number.

17. a. Incorrect. Erythema is the term for reddening of the skin caused by radiation damage.

b. Incorrect. Erythema is the term for reddening of the skin caused by radiation damage.

c. Correct. Erythema is the term for reddening of the skin caused by radiation damage.

d. Incorrect. Erythema is the term for reddening of the skin caused by radiation damage.

18. a. Incorrect. Radiation cataractogenesis, or radiation-induced cataracts, exhibit a nonlinear, threshold dose-response relationship.

b. Correct. Radiation cataractogenesis, or radiation-induced cataracts, exhibit a nonlinear, threshold dose-response relationship.

c. Incorrect. Radiation cataractogenesis, or radiation-induced cataracts, exhibit a nonlinear, threshold dose-response relationship. The threshold is approximately 200 rad.

d. Incorrect. Radiation cataractogenesis, or radiation-induced cataracts, exhibit a nonlinear, threshold dose-response relationship. The latent period is approximately 15 years.

19. a. Correct. The study of populations is the definition of epidemiology.

b. Incorrect. The study of populations is the definition of epidemiology.

c. Incorrect. The study of populations is the definition of epidemiology.

d. Incorrect. The study of populations is the definition of epidemiology.

20. a. Incorrect. The relative risk for developing radiation-induced leukemia is approximately 1.5.

b. Incorrect. The relative risk for developing radiation-induced leukemia is approximately 1.5.

c. Incorrect. The relative risk for developing radiation-induced leukemia is approximately 1.5.

d. Correct. The relative risk for developing radiation-induced leukemia is approximately 1.5.

21. a. Incorrect. The absolute risk factor for radiation-induced breast cancer is approximately 6 cases/10^6/rad/year.

b. Incorrect. The absolute risk factor for radiation-induced breast cancer is approximately 6 cases/10^6/rad/year.

c. Correct. The absolute risk factor for radiation-induced breast cancer is approximately 6 cases/10^6/rad/year.

d. Incorrect. The absolute risk factor for radiation-induced breast cancer is approximately 6 cases/10^6/rad/year.

22. a. Incorrect. Radiation-induced lung cancer follows a linear, nonthreshold dose-response relationship.

b. Correct. Radiation-induced lung cancer follows a linear, nonthreshold dose-response relationship.

c. Incorrect. Radiation-induced lung cancer follows a linear, nonthreshold dose-response relationship.

d. Incorrect. Radiation-induced lung cancer follows a linear, nonthreshold dose-response relationship.

23. a. Incorrect. The approximate latent period for radiation-induced leukemia is 4-7 years.

b. Incorrect. The approximate latent period for radiation-induced leukemia is 4–7 years.

c. Correct. The approximate latent period for radiation-induced leukemia is 4–7 years.

d. Incorrect. The approximate latent period for radiation-induced leukemia is 4–7 years.

24. a. Correct. American radiographers have not shown an increased incidence of radiation-induced leukemia.

b. Incorrect. American radiologists have shown an increased incidence of radiation-induced leukemia.

c. Incorrect. Atomic bomb survivors have shown an increased incidence of radiation-induced leukemia.

d. Incorrect. Radiotherapy patients have shown an increased incidence of radiation-induced leukemia.

25. a. Correct. Radiation-induced skin cancer follows a threshold dose-response relationship.

b. Incorrect. Radiation-induced bone cancer follows a nonthreshold dose-response relationship.

c. Incorrect. Radiation-induced lung cancer follows a nonthreshold dose-response relationship.

d. Incorrect. Radiation-induced breast cancer follows a nonthreshold dose-response relationship.

26. a. Incorrect. If in utero radiation is received during the organogenesis stage, the most likely result would be the presence of congenital abnormalities.

b. Incorrect. If in utero radiation is received during the organogenesis stage, the most likely result would be the presence of congenital abnormalities.

c. Incorrect. If in utero radiation is received during the organogenesis stage, the most likely result would be the presence of congenital abnormalities.

d. Correct. If in utero radiation is received during the organogenesis stage, the most likely result would be the presence of congenital abnormalities.

27. a. Incorrect. Stochastic is the term used to describe radiation damage which increases the probability of causing a late effect but will not increase the severity of the effect.

b. Incorrect. Stochastic is the term used to describe radiation damage which increases the probability of causing a late effect but will not increase the severity of the effect.

c. Correct. Stochastic is the term used to describe radiation damage which increases the probability of causing a late effect but will not increase the severity of the effect.

d. Incorrect. Stochastic is the term used to describe radiation damage which increases the probability of causing a late effect but will not increase the severity of the effect.

28. a. Correct. Doubling dose is the term used to describe the dose of radiation which will increase the number of mutations by a factor of two.

b. Incorrect. Doubling dose is the term used to describe the dose of radiation which will increase the number of mutations by a factor of two.

c. Incorrect. Doubling dose is the term used to describe the dose of radiation which will increase the number of mutations by a factor of two.

d. Incorrect. Doubling dose is the term used to describe the dose of radiation which will increase the number of mutations by a factor of two.

29. a. Incorrect. The approximate dose necessary to cause permanent male sterility is 500 rad.

b. Incorrect. The approximate dose necessary to cause permanent male sterility is 500 rad.

c. Incorrect. The approximate dose necessary to cause permanent male sterility is 500 rad.

d. Correct. The approximate dose necessary to cause permanent male sterility is 500 rad.

30. a. Incorrect. An acute whole body exposure of 5,000 rad would result in the central nervous system syndrome.

b. Incorrect. An acute whole body exposure of 5,000 rad would result in the central nervous system syndrome.

c. Correct. An acute whole body exposure of 5,000 rad would result in the central nervous system syndrome.

31. a. Incorrect. Desquamation is the term used for ulceration, necrosis, and loss of skin cells caused by irradiation.

b. Incorrect. Desquamation is the term used for ulceration, necrosis, and loss of skin cells caused by irradiation.

c. Correct. Desquamation is the term used for ulceration, necrosis, and loss of skin cells caused by irradiation.

32. a. Correct. When compared with the types of patients listed, the fetus is considered the most radiosensitive.

b. Incorrect. When compared with the types of patients listed, the fetus is considered the most radiosensitive.

c. Incorrect. When compared with the types of patients listed, the fetus is considered the most radiosensitive.

d. Incorrect. When compared with the types of patients listed, the fetus is considered the most radiosensitive.

33. a. Incorrect. The annual dose equivalent limit for the fetus of an occupational worker is 0.5 rem.

b. Incorrect. The annual dose equivalent limit for the fetus of an occupational worker is 0.5 rem.

c. Correct. The annual dose equivalent limit for the fetus of an occupational worker is 0.5 rem.

d. Incorrect. The annual dose equivalent limit for the fetus of an occupational worker is 0.5 rem.

Chapter 6

Key Terms Review

s Agreement states
r ALARA
q Bone marrow dose
t Controlled area
p Cumulative timing device
x Deadman type
w Dose
u Early effect of radiation
o Dosimetry
n Effective dose-equivalent limit
m Entrance skin exposure (ESE)
l Film badge
k Gonadal dose
gg Intermittent fluoroscopy
j Inverse square law
i Ionization chamber

ff Joint Commission on the Accreditation of Healthcare Organizations (JCAHO)

y Late effect of radiation

h Lead equivalent

aa Personnel dosimeters

z Pocket dosimeters

f Primary protective barrier

v Protective tube housing

e Radiation protection

d Scattered radiation

c Secondary protective barrier

cc Skin dose

ee Thermoluminescent dosimeters (TLD)

g Time of occupancy

dd Uncontrolled area

bb Use (U)

b Useful beam

a Workload (W)

Review

1. Early effects of radiation include the following: hematologic depression, skin erythema, epilation, chromosome damage, gonad dysfunction, death.

2. Do not wear when seen as a patient.
Store in an area free from excess heat and high humidity.
Wear only for the designated time period.
To be worn only by person assigned to that badge.

3. TLDs: are nearly tissue equivalent, can be worn for 3 months, are reusable, are extremely accurate, and are more sensitive than film badges.

4. Pocket dosimeters provide an immediate response.

5. An institution's radiation monitoring program guidelines are established by the radiation safety officer (RSO).

6. Present-day radiation protection guidelines are established using the ALARA philosophy, that is, keep exposures *As Low As Reasonably Achievable*.

7. Effective dose-equivalent limit during any 13-week period is 3 rem.

8. A pregnant radiographer should by no means be removed from her duties. The RSO should review her previous exposure history, her work environment should be analyzed to estimate the likelihood of receiving exposures that would be greater than 0.5 rem limit, and her work routines should be altered to lessen the chance of risks.

9. Formula used for calculating cumulative whole-body dose-equivalent limit is:

Age in years × 1 rem.

10. PA chest gonadal dose = <1 mrad.

Lumbar spine gonadal dose = 225 mrad.

11. Diagnostic radiology personnel may receive an annual effective dose-equivalent limit of ________________ for whole-body occupational exposure during routine operations.

a. Incorrect. The annual effective dose equivalent limit for whole-body occupational exposure is 5 rem (50 mSv).

b. Incorrect. The annual effective dose equivalent limit for whole-body occupational exposure is 5 rem (50 mSv).

c. Correct. The annual effective dose equivalent limit for whole-body occupational exposure is 5 rem (50 mSv).

d. Incorrect. The annual effective dose equivalent limit for whole-body occupational exposure is 5 rem (50 mSv).

12. Primary protective barriers must consist of _____ inch lead.

a. Correct. Primary protective barriers consist of 1/16 inch Pb.

b. Incorrect. Primary protective barriers consist of 1/16 inch Pb.

c. Incorrect. Primary protective barriers consist of 1/16 inch Pb.

d. Incorrect. Primary protective barriers consist of 1/16 inch Pb. Secondary protective barriers consist of 1/32 inch Pb.

13. Shielding for an uncontrolled area maintains the exposure remains below _______ mR/week.

a. Correct. Uncontrolled area design limits require barriers to reduce the exposure rate to less than 10 mR/week.

b. Incorrect. Uncontrolled area design limits require barriers to reduce the exposure rate to less than 10 mR/week.

c. Incorrect. Uncontrolled area design limits require barriers to reduce the exposure rate to less than 10 mR/week. Controlled area design limits require barriers to reduce the exposure rate to less than 100 mR/week.

d. Incorrect. Uncontrolled area design limits require barriers to reduce the exposure rate to less than 10 mR/week.

14. The exposure switch cord on a mobile unit must be long enough to allow the radiographer to stand back at least ______ feet during the exposure.

a. Incorrect. The exposure switch of the mobile unit must allow the radiographer to remain at least 6 feet from the x-ray tube during the exposure.

b. Incorrect. The exposure switch of the mobile unit must allow the radiographer to remain at least 6 feet from the x-ray tube during the exposure.

c. Correct. The exposure switch of the mobile unit must allow the radiographer to remain at least 6 feet from the x-ray tube during the exposure.

d. Incorrect. The exposure switch of the mobile unit must allow the radiographer to remain at least 6 feet from the x-ray tube during the exposure.

15. A Bucky slot cover should be at least ______________.

a. Incorrect. The Bucky slot cover must be at least 0.25 mm Pb equivalent.

b. Correct. The Bucky slot cover must be at least 0.25 mm Pb equivalent.

c. Incorrect. The Bucky slot cover must be at least 0.25 mm Pb equivalent.

d. Incorrect. The Bucky slot cover must be at least 0.25 mm Pb equivalent.

16. Protective lead gloves must have a minimum lead equivalency of at least ______ mm lead.

a. Incorrect. Lead gloves must have a minimum lead equivalency of at least 0.25 mm.

b. Correct. Lead gloves must have a minimum lead equivalency of at least 0.25 mm.

c. Incorrect. Lead gloves must have a minimum lead equivalency of at least 0.25 mm.

d. Incorrect. Lead gloves must have a minimum lead equivalency of at least 0.25 mm.

17. Which of the following people should be asked to hold a patient who needs physical support during the exposure?

a. Incorrect. A nonoccupational person should be asked to hold a patient when necessary.

b. Correct. A nonoccupational person should be asked to hold a patient when necessary.

c. Incorrect. A nonoccupational person should be asked to hold a patient when necessary.

d. Incorrect. A nonoccupational person should be asked to hold a patient when necessary.

Chapter 7

Key Terms Review

o Added filtration
p Aperture diaphragm
q Beam limitation device
r Cones
s Cumulative timing device
g Flat contact shield
h Gonad shielding
i Half-value layer (HVL)
j Image intensification flouroscopy
k Inherent filtration
l Intensifying screen
m Intermittent flouroscopy
n Involuntary motion
a Shadow shield
b Shaped contact shield
c Source-image receptor distance (SID)
d Total filtration
e Variable-aperture collimator
f Voluntary motion

Review

1. a. Correct. The use of immobilization devices tends to decrease the possibility of voluntary patient motion.

b. Incorrect. Reducing radiation exposure times would be used to overcome involuntary motion.

c. Incorrect. Communication is vital in order to prepare the patient for what is expected during the exam.

2. a. Incorrect. This device is limited in use since it has a hole of a designated size and shape cut in its center.

b. Incorrect. This device is limited in use since it has a predetermined circular size and shape.

c. Incorrect. This device is limited in use since it has a predetermined circular size and shape.

d. Correct. This device is the most versatile, since it may be adjusted so that a variety of rectangular shapes may be selected.

3. a. Incorrect. A decrease in short wavelength radiation would result in decreasing patient skin dose.

b. Incorrect. Increasing beam hardness would decrease patient skin dose.

c. Correct. Increasing beam hardness would reduce patient skin dose.

d. Incorrect. Decreasing beam hardness would increase patient skin dose.

4. a. Incorrect. Shadow shields that attach to the tube head are particularly useful for sterile fields.

b. Incorrect. Flat contact shields are useful for simple recumbent studies, but when the exam necessitates that oblique, lateral, or erect projections be obtained, they become less efficient.

c. Correct. Shaped contact shields are best because they enclose the male reproductive organs, remaining in position in oblique, lateral, and erect positions.

5. a. Correct. Use of high kVp, low mAs, and increased filtration results in the lowest patient dose possible.

b. Incorrect. Use of low kVp, high mAs, and decreased filtration results in high patient dose.

c. Incorrect. Use of high kVp results in low patient dose. Use of high mAs and decreased filtration results in high patient dose.

d. Incorrect. Use of low kVp results in high patient dose.

6. a. Correct. Patient dose decreases when using a combination of high-speed film with high-speed screens.

b. Incorrect. Patient dose would decrease if rare-earth screens were used as compared to using calcium tungstate screens.

c. Incorrect. Low kVp techniques increase patient dose.

d. Incorrect. Use of non-screen film increases patient dose.

7. a. Incorrect. Having to repeat a radiograph results in the patient receiving twice the normal exposure.

b. Correct. A repeat radiograph results in twice the exposure to the patient.

8. a. Correct. Image intensification increases the brightness of the image 7,000 times as compared with conventional fluoroscopy.

b. Incorrect. Since the image brightness is intensified 7,000 times as compared with conventional fluoroscopy, the radiologist does not have to undergo dark adaptation.

c. Incorrect. Image intensified fluoroscopy uses 1.5–2 mA, as compared with conventional fluoroscopy, which uses 3–5 mA. The use of the lower mA decreases patient dose.

d. Incorrect. Image intensification increases the brightness of the image 7,000 times as compared with conventional fluoroscopy.

9. a. Incorrect. Source-tabletop distance is 15 in. (38 cm) for fixed, 12 in. (30 cm) for mobile fluoroscopes.

b. Correct. Source-tabletop distance is 15 in. (38 cm) for fixed, 12 in. (30 cm) for mobile fluoroscopes.

c. Incorrect. Source-tabletop distance is 15 in. (38 cm) for fixed, 12 in. (30 cm) for mobile fluoroscopes.

d. Incorrect. Source-tabletop distance is 15 in. (38 cm) for fixed, 12 in. (30 cm) for mobile fluoroscopes. 149-173 section 3

SECTION III SELF-ASSESSMENT ANSWERS

1. a. 1 rem. Incorrect. The effective dose-equivalent limit for the lens of the eye is 15 rem.

b. 5 rem. Incorrect. The effective dose-equivalent limit for the lens of the eye is 15 rem. 5 rem is the annual effective dose-equivalent limit for whole-body occupational exposure.

c. 10 rem. Incorrect. The effective dose-equivalent limit for he lens of the eye is 15 rem.

d. 15 rem. Correct. The effective dose-equivalent limit for the lens of the eye is 15 rem.

2. a. 0.1 rem. Correct. The effective dose-equivalent limit for student radiographers under the age of 18 is 0.1 rem.

b. 0.5 rem. Incorrect. The effective dose-equivalent limit for student radiographers under the age of 18 is 0.1 rem. 0.5 rem is the total effective dose-equivalent limit for a pregnant radiographer.

c. 1 rem. Incorrect. The effective dose-equivalent limit for student radiographers under the age of 18 is 0.1 rem.

d. 5 rem. Incorrect. The effective dose-equivalent limit for student radiographers under the age of 18 is 0.1 rem. 5 rem is the annual dose-equivalent limit for whole-body occupational exposure.

3. a. 0.1 rem. Incorrect. The total effective dose-equivalent limit for a pregnant radiographer is 0.5 rem. 0.1 rem is the effective dose-equivalent limit for a student radiographer under the age of 18.

b. 0.5 rem. Correct. The total effective dose-equivalent limit for a pregnant radiographer is 0.5 rem.

c. 1 rem. Incorrect. The total effective dose-equivalent limit for a pregnant radiographer is 0.5 rem.

d. 5 rem. Incorrect. The total effective dose-equivalent limit for a pregnant radiographer is 0.5 rem.

4. a. 1/64 inch. Incorrect. Secondary protective barriers should consist of 1/32 inch Pb.

b. 1/32 inch. Correct. Secondary protective barriers should consist of 1/32 inch Pb.

c. 1/16 inch. Incorrect. Secondary protective barriers should consist of 1/32 inch Pb.

d. 1/4 inch. Incorrect. Secondary protective barriers should consist of 1/32 inch Pb.

5. a. 3 feet. Incorrect. Primary protective barriers must be 7 feet tall.

b. 4 feet. Incorrect. Primary protective barriers must be 7 feet tall.

c. 7 feet. Correct. Primary protective barriers must be 7 feet tall.

d. 10 feet. Incorrect. Primary protective barriers must be 7 feet tall.

6. a. 1 mR/hr. Incorrect. The protective tube housing serves to reduce leakage radiation to less than 100 mR/hr at a distance of 3 feet from the tube.

b. 10 mR/hr. Incorrect. The protective tube housing serves to reduce leakage radiation to less than 100 mR/hr at a distance of 3 feet from the tube.

c. 100 mR/hr. Correct. The protective tube housing serves to reduce leakage radiation to less than 100 mR/hr at a distance of 3 feet from the tube.

d. 1,000 mR/hr. Incorrect. The protective tube housing serves to reduce leakage radiation to less than 100 mR/hr at a distance of 3 feet from the tube.

7. a. 0.001%. Incorrect. At 1 meter from the patient the beam intensity is reduced by a factor of 1,000, to approximately 0.1% of the original beam intensity.

b. 0.01%. Incorrect. At 1 meter from the patient the beam intensity is reduced by a factor of 1,000, to approximately 0.1% of the original beam intensity.

c. 0.1%. Correct. At 1 meter from the patient the beam intensity is reduced by a factor of 1,000, to approximately 0.1% of the original beam intensity.

d. 1%. Incorrect. At 1 meter from the patient the beam intensity if reduced by a factor of 1,000, to approximately 0.1% of the original beam intensity.

8. a. Occupancy factor. Incorrect. Occupancy factor is a consideration of who occupies a particular area—radiation versus non-radiation workers.

b. Use factor. Correct. Use factor is a description of the percentage of time which the x-ray beam is energized and directed toward a particular wall.

c. Workload. Incorrect. Workload is calculated by the number of x-ray exposures made weekly.

9. a. 0.12 mm. Incorrect. The lead equivalency of an apron must be 0.5 mm Pb.

b. 0.25 mm. Incorrect. The lead equivalency of an apron must be 0.5 mm Pb.

c. 0.5 mm. Correct. The lead equivalency of an apron must be 0.5 mm Pb.

d. 1 mm. Incorrect. The lead equivalency of an apron must be 0.5 mm Pb.

10. a. Abort fluoroscopic exposure after 5 minutes. Incorrect. Fluoroscopic exposure switches must be of the deadman type. The fluoroscopic timer will abort fluoro exposure after 5 minutes.

b. Be at the end of a six foot long expandable cord. Incorrect. Fluoroscopic exposure switches must be of the deadman type.

c. Make an audible sound during fluoroscopic exposure. Incorrect. Fluoroscopic exposure switches must be of the deadman type. The fluoroscopic timer will make an audible sound after 5 minutes of fluoroscopic exposure time.

d. Be of the deadman type. Correct. Fluoroscopic exposure switches must be of the dead-man type.

11. a. 6 inches. Incorrect. The source-to-tabletop distance in mobile fluoroscopy must not be less than 12 inches.

b. 12 inches. Correct. The source-to-tabletop distance in mobile fluoroscopy must not be less than 12 inches.

c. 15 inches. Incorrect. The source-top-tabletop distance in mobile fluoroscopy must not be less than 12 inches. The source-to-tabletop distance in fixed fluoroscopy must not be less than 15 inches.

d. 20 inches. Incorrect. The source-to-tabletop distance in mobile fluoroscopy must not be less than 12 inches.

12. a. 0.025 mm. Incorrect. Fluoroscopy lead curtain must be at least 0.25 mm Pb equivalency.

b. 0.25 mm. Correct. Fluoroscopy lead curtains must be at least 0.25 mm Pb equivalency.

c. 2.5 mm. Incorrect. Fluoroscopy lead curtains must be at least 0.25 mm Pb equivalency.

d. 25 mm. Incorrect. Fluoroscopy lead curtains must be at least 0.25 mm Pb equivalency.

13. a. 1 R/min. Incorrect. Tabletop x-ray intensity during fluoroscopy must not exceed 10 R/min.

b. 5 R/min. Incorrect. Tabletop x-ray intensity during fluoroscopy must not exceed 10 R/min.

c. 10 R/min. Correct. Tabletop x-ray intensity during fluoroscopy must not exceed 10 R/min.

d. 100 R/min. Incorrect. Tabletop x-ray intensity during fluoroscopy must not exceed 10 R/min.

14. a. 1 mR/min. Incorrect. Utilizing the inverse square law, if the distance from a source is doubled, the dose/intensity will be reduced by a factor of 4. The inverse square law is:

$$\frac{I_1}{I_2} = \frac{d_2^2}{d_1^2}$$

Substituting known values:

$$\frac{8\ \text{mR/min}}{I_2} = \frac{6^2}{3^2}$$

$$\frac{8}{I_2} = \frac{36}{9}$$

Cross multiply factors.

$36I_2 = 72$

$I_2 = 2$

b. 2 mR/min. Correct. Utilizing the inverse square law, if the distance from a source is doubled, the dose/intensity will be reduced by a factor of 4. The inverse square law is:

$$\frac{I_1}{I_2} = \frac{d_2^2}{d_1^2}$$

Substituting known values:

$$\frac{8\ \text{mR/min}}{I_2} = \frac{6^2}{3^2}$$

$$\frac{8}{I_2} = \frac{36}{9}$$

Cross multiply factors.

$36I_2 = 72$

$I_2 = 2$

c. 16 mR/min. Incorrect. Utilizing the inverse square law, if the distance from a source is doubled, the dose/intensity will be reduced by a factor of 4. The inverse square law is:

$$\frac{I_1}{I_2} = \frac{d_2^2}{d_1^2}$$

Substituting known values:

$$\frac{8\ \text{mR/min}}{I_2} = \frac{6^2}{3^2}$$

$$\frac{8}{I_2} = \frac{36}{9}$$

Cross multiply factors.

$36I_2 = 72$

$I_2 = 2$

d. 32 mR/min. Incorrect. Utilizing the inverse square law, if the distance from a source is doubled, the dose/intensity will be reduced by a factor of 4. The inverse square law is:

$$\frac{I_1}{I_2} = \frac{d_2^2}{d_1^2}$$

Substituting known values:

$$\frac{8\ \text{mR/min}}{I_2} = \frac{6^2}{3^2}$$

$$\frac{8}{I_2} = \frac{36}{9}$$

Cross multiply factors.

$36I_2 = 72$

$I_2 = 2$

15. a. 1 ft. Incorrect. The inverse square law is:

$$\frac{I_1}{I_2} = \frac{d_2^2}{d_1^2}$$

Substituting known values:

$$\frac{40\ \text{R/hr}}{10\ \text{R/hr}} = \frac{d_2^2}{3^2}$$

$$\frac{40}{10} = \frac{d_2^2}{9}$$

Cross multiply factors.

$10d_2^2 = 360$

$d_2^2 = 36$

$d_2 = 6$

b. 2 ft. Incorrect. The inverse square law is:

$$\frac{I_1}{I_2} = \frac{d_2^2}{d_1^2}$$

Substituting known values:

$$\frac{40 \text{ R/hr}}{10 \text{ R/hr}} = \frac{d_2^2}{3^2}$$

$$\frac{40}{10} = \frac{d_2^2}{9}$$

Cross multiply factors.

$10d_2^2 = 360$

$d_2^2 = 36$

$d_2 = 6$

c. 6 ft. Correct. The inverse square law is:

$$\frac{I_1}{I_2} = \frac{d_2^2}{d_1^2}$$

Substituting known values:

$$\frac{40 \text{ R/hr}}{10 \text{ R/hr}} = \frac{d_2^2}{3^2}$$

$$\frac{40}{10} = \frac{d_2^2}{9}$$

Cross multiply factors.

$10d_2^2 = 360$

$d_2^2 = 36$

$d_2 = 6$

d. 12 ft. Incorrect. The inverse square law is:

$$\frac{I_1}{I_2} = \frac{d_2^2}{d_1^2}$$

Substituting known values:

$$\frac{40 \text{ R/hr}}{10 \text{ R/hr}} = \frac{d_2^2}{3^2}$$

$$\frac{40}{10} = \frac{d_2^2}{9}$$

Cross multiply factors.

$10d_2^2 = 360$

$d_2^2 = 36$

$d_2 = 6$

16. a. 0.01. Incorrect. Monitoring of personnel is required when personnel receive 10% of the annual effective dose-equivalent limit.

b. 1. Incorrect. Monitoring of personnel is required when personnel receive 10% of the annual effective dose-equivalent limit.

c. 5. Incorrect. Monitoring of personnel is required when personnel receive 10% of the annual effective dose-equivalent limit.

d. 10. Correct. Monitoring of personnel is required when personnel receive 10% of the annual effective dose-equivalent limit.

17. a. 15. Incorrect. The cumulative whole-body effective dose-equivalent limit is calculated by multiplying one's age in years times 1 rem. 45 x 1 rem = 45 rem.

b. 45. Correct. The cumulative whole-body effective dose-equivalent limit is calculated by multiplying one's age in years times 1 rem. 45 x 1 rem = 45 rem.

c. 35. Incorrect. The cumulative whole-body effective dose-equivalent limit is calculated by multiplying one's age in years times 1 rem. 45 x 1 rem = 45 rem.

d. 25. Incorrect. The cumulative whole-body effective dose-equivalent limit is calculated by multiplying one's age in years times 1 rem. 45 x 1 rem = 45 rem.

18. a. 10. Correct. Thyroid shields reduce radiation dose by a factor of approximately 10.

b. 7. Incorrect. Thyroid shields reduce radiation dose by a factor of approximately 10.

c. 4. Incorrect. Thyroid shields reduce radiation dose by a factor of approximately 10.

d. 1. Incorrect. Thyroid shields reduce radiation dose by a factor of approximately 10.

19. a. 150 mAs. Incorrect. If mAs is increased, intensity also increases. The formula for mAs and intensity is:

$$\frac{mAs_1}{mAs_2} = \frac{I_1}{I_2}$$

$$\frac{20}{40} = \frac{300}{x} \quad 20x = 12{,}000 \quad x = 600.$$

b. 300 mAs. Incorrect. If mAs is increased, intensity also increases. The formula for mAs and intensity is:

$$\frac{mAs_1}{mAs_2} = \frac{I_1}{I_2}$$

$$\frac{20}{40} = \frac{300}{x} \quad 20x = 12{,}000 \quad x = 600.$$

c. 600 mAs. Correct. If mAs is increased, intensity also increases. The formula for mAs and intensity is:

$$\frac{mAs_1}{mAs_2} = \frac{I_1}{I_2}$$

$$\frac{20}{40} = \frac{300}{x} \quad 20x = 12{,}000 \quad x = 600.$$

d. 900 mAs. Incorrect. If mAs is increased, intensity

$$\frac{mAs_1}{mAs_2} = \frac{I_1}{I_2}$$

$$\frac{20}{40} = \frac{300}{x} \quad 20x = 12{,}000 \quad x = 600.$$

20. a. 100. Incorrect. Increasing kVp should also increase intensity. Changing kVp affects radiation intensity by the square of the ratios of the kVps. The following is the formula for kVp and radiation intensity:

$$\frac{(kVp_1)^2}{(kVp_2)^2} = \frac{I_1}{I_2}$$

$$\frac{(70)^2}{(100)^2} = \frac{200}{x}$$

$$\frac{49}{100} = \frac{200}{x}$$

$49x = 20{,}000$

$x = 408$

b. 300. Incorrect. Increasing kVp should also increase intensity. Changing kVp affects radiation intensity by the square of the ratios of the kVps. The following is the formula for kVp and radiation intensity:

$$\frac{(kVp_1)^2}{(kVp_2)^2} = \frac{I_1}{I_2}$$

$$\frac{(70)^2}{(100)^2} = \frac{200}{x}$$

$$\frac{49}{100} = \frac{200}{x}$$

$49x = 20{,}000$

$x = 408$

c. 408. Correct. Increasing kVp should also increase intensity. Changing kVp affects radiation intensity by the square of the ratios of the kVps. The following is the formula for kVp and radiation intensity:

$$\frac{(kVp_1)^2}{(kVp_2)^2} = \frac{I_1}{I_2}$$

$$\frac{(70)^2}{(100)^2} = \frac{200}{x}$$

$$\frac{49}{100} = \frac{200}{x}$$

$49x = 20{,}000$

$x = 408$

d. 808. Incorrect. Increasing kVp should also increase intensity. Changing kVp affects radiation intensity by the square of the ratios of the kVps. The following is the formula for kVp and radiation intensity:

$$\frac{(kVp_1)^2}{(kVp_2)^2} = \frac{I_1}{I_2}$$

$$\frac{(70)^2}{(100)^2} = \frac{200}{x}$$

$$\frac{49}{100} = \frac{200}{x}$$

$49x = 20{,}000$

$x = 408$

21. a. True. Incorrect. Voluntary patient motion is reduced by using immobilization.

b. False. Correct. Shorter exposure times are used to compensate for involuntary patient motion

22. a. 2%. Correct. Federal guidelines set forth for patient protection state that the collimator light and actual irradiated area must be accurate to within 2% of the SID.

b. 5%. Incorrect. Federal guidelines set forth for patient protection state that the collimator light and actual irradiated area must be accurate to within 2% of the SID.

c. 10% Incorrect. Federal guidelines set forth for patient protection state that the collimator light and actual irradiated area must be accurate to within 2% of the SID.

d. 22%. Incorrect. Federal guidelines set forth for patient protection state that the collimator light and actual irradiated area must be accurate to within 2% of the SID.

23. a. Shorter, bigger. Incorrect. The greatest beam limitation occurs with the longest cone and smallest diameter available.

b. Shorter, smaller. Incorrect. The greatest beam limitation occurs with the longest cone and smallest diameter available.

c. Longer, bigger. Incorrect. The greatest beam limitation occurs with the longest cone and smallest diameter available.

d. Longer, smaller. Correct. The greatest beam limitation occurs with the longest cone and smallest diameter available.

24. a. Short wavelength radiation. Incorrect. The function of filtration is to remove long wavelength radiation.

b. Long wavelength radiation. Correct. The function of filtration is to remove long wavelength radiation.

c. Secondary radiation. Incorrect. The function of filtration is to remove long wavelength radiation.

d. Scattered radiation. Incorrect. The function of filtration is to remove long wavelength radiation.

25. a. 1.4 mm Al. Incorrect. When using over 70 kVp, 2.5 mm Al is required.

b. 2.5 mm Al. Correct. When using over 70 kVp, 2.5 mm Al is required.

c. 3.0 mm Al. Incorrect. When using over 70 kVp, 2.5 mm Al is required.

d. 4.0 mm Al. Incorrect. When using over 70 kVp, 2.5 mm Al is required.

26. a. Flat contact shields. Incorrect. Shadow shields are useful when encountering sterile fields.

b. Shadow shields. Correct. Shadow shields are useful when encountering sterile fields.

c. Shaped contact shields. Incorrect. Shadow shields are useful when encountering sterile fields.

d. Lead sheet. Incorrect. Shadow shields are useful when encountering sterile fields.

27. a. 50 mAs, 90 kVp. Correct. The lowest mAs and highest kVp possible should be selected to provide the least amount of patient exposure.

b. 100 mAs, 90 kVp. Incorrect. The lowest mAs and highest kVp possible should be selected in order to provide the least amount of patient exposure.

c. 200 mAs, 50 kVp. Incorrect. The lowest mAs and highest kVp possible should be selected in order to provide the least amount of patient exposure.

d. 400 mAs, 50 kVp. Incorrect. The lowest mAs and highest kVp possible should be selected in order to provide the least amount of patient exposure.

28. a. Use non-screen film. Incorrect. Use of non-screen film increases patient exposure.

b. Use rare-earth screens. Correct. Use of rare-earth screens significantly reduces patient exposure.

c. Decrease kVp. Incorrect. Decreasing kVp will increase patient exposure.

d. Increase mAs. Incorrect. Increasing mAs will increase patient exposure.

29. a. Disregard communication. Incorrect. Communication is vital in order to prepare the patient for what is expected during the exam.

b. Eliminate voluntary patient motion using short exposure times. Incorrect. Voluntary patient motion is eliminated by using immobilization devices.

c. Eliminate involuntary patient motion using immobilization devices. Incorrect. Involuntary patient motion is eliminated by using short exposure times.

d. Eliminate voluntary patient motion using immobilization devices. Correct. Immobilization devices eliminate voluntary patient motion, thus reducing the chance of a repeat exam.

30. a. 1 R/min. Incorrect. Tabletop exposure rate shall not exceed 10 R/min during fluoroscopy.

b. 5 R/min. Incorrect. Tabletop exposure rate shall not exceed 10 R/min during fluoroscopy.

c. 7 R/min. Incorrect. Tabletop exposure rate shall not exceed 10 R/min during fluoroscopy.

d. 10 R/min. Correct. Tabletop exposure rate shall not exceed 10 R/min during fluoroscopy.

31. a. 1 minute. Incorrect. The fluoroscopic timer must sound an audible alarm after 5 minutes.

b. 2 minutes. Incorrect. The fluoroscopic timer must sound an audible alarm after 5 minutes.

c. 3 minutes. Incorrect. The fluoroscopic timer must sound an audible alarm after 5 minutes.

d. 5 minutes. Correct. The fluoroscopic timer must sound an audible alarm after 5 minutes.

32. a. Collimator. Incorrect. Collimators are the most versatile type of beam limitation device.

b. Cone. Incorrect. Cones are a type of beam limitation device which attach to the x-ray tube head.

c. Filter. Correct. Filters are not beam limitation devices. Filter are inserted into the path of the x-ray beam to absorb low energy photons.

d. Aperture diaphragm. Incorrect. Aperture diaphragms are beam limitation devices which attach to the x-ray tube head.

33. a. Minimum of 7 inches. Incorrect. Fixed fluoroscopic equipment must provide at least 15 inches source-to-tabletop distance for the protection of the patient.

b. Minimum of 12 inches. Incorrect. Fixed fluoroscopic equipment must provide at least 15 inches source-to-tabletop distance for the protection of the patient.

c. Minimum of 15 inches. Correct. Fixed fluoroscopic equipment must provide at least 15 inches source-to-tabletop distance for the protection of the patient.

d. A maximum of 12 inches. Incorrect. Fixed fluoroscopic equipment must provide at least 15 inches source-to-tabletop distance for the protection of the patient.

34. a. Decrease intensity by 50%. Correct. HVL is the thickness of absorbing material necessary to reduce the x-ray intensity to half its original value.

b. Decrease intensity by 25%. Incorrect. HVL is the thickness of absorbing material necessary to reduce the x-ray intensity to half its original value.

c. Increase intensity by 50%. Incorrect. HVL is the thickness of absorbing material necessary to reduce the x-ray intensity to half its original value.

d. Increase intensity by 25%. Incorrect. HVL is the thickness of absorbing material necessary to reduce the x-ray intensity to half its original value.

35. a. 4 inches. Incorrect. Source-to-table distance in mobile fluoroscopy shall not be less than 12 inches.

b. 9 inches. Incorrect. Source-to-tabletop distance in mobile fluoroscopy shall not be less than 12 inches.

c. 12 inches. Correct. Source-to-tabletop distance in mobile fluoroscopy shall not be less than 12 inches.

d. 15 inches. Incorrect. Source-to-tabletop distance in mobile fluoroscopy shall not be less than 12 inches.